SpringerBriefs in Electrical and Computer Engineering

More information about this series at http://www.springer.com/series/10059

Lu Yang • Wei Zhang

Interference Coordination for 5G Cellular Networks

 Springer

Lu Yang
The University of New South Wales
Sydney, Australia

Wei Zhang
The University of New South Wales
Sydney, Australia

ISSN 2191-8112   ISSN 2191-8120   (electronic)
SpringerBriefs in Electrical and Computer Engineering
ISBN 978-3-319-24721-2   ISBN 978-3-319-24723-6   (eBook)
DOI 10.1007/978-3-319-24723-6

Library of Congress Control Number: 2015951203

Springer Cham Heidelberg New York Dordrecht London

Printed on acid-free paper

Springer International Publishing AG Switzerland is part of Springer Science+Business Media (www.springer.com)

Preface

With the 4G cellular networks reaching saturation, the 5G networks have recently attracted growing attention and research efforts in both academic and industry communities. The 5G cellular networks are featured as very-high-density base stations and wireless devices and integration of heterogeneous networks. As a result, advanced interference management schemes are critical for achieving the desired network throughput. Due to the high density of wireless devices, users located in the cell boundary area probably share the same channel resources with other users in the same area associated with neighboring cells, thereby causing severe interference to each other. In addition, the heterogeneous networks will add complexity in dealing with the mutual interference. This Springer Brief presents interference coordination techniques for future 5G cellular networks. Starting with an overview of existing interference management techniques, this Brief focuses on practical interference coordination schemes based on beamforming and user scheduling. The proposed schemes aim to deal with the intercell interference in multicell MIMO networks, cross-tier interference in device-to-device communications underlaying cellular networks, and inter-network interference in cognitive radio networks. The performances of the proposed schemes are evaluated both analytically and numerically in terms of several performance parameters, including the sum rate, multiplexing gain, and outage probability of the networks. The results show that the proposed schemes can significantly reduce the effect of interference and improve the quality of service of the networks.

Sydney, Australia
Sydney, Australia

Lu Yang
Wei Zhang

v

Contents

Acronyms

5G	Fifth generation
BS	Base station
ICI	Intercell interference
CSI	Channel state information
MIMO	Multiple-input multiple-output
D2D	Device to device
M2M	Machine to machine
CR	Cognitive radio
OCI	Other-cell interference
TDMA	Time-division multiple access
OFDMA	Orthogonal frequency-division multiple access
SINR	Signal-to-interference-and-noise ratio
BER	Bit error rate
DoF	Degrees of freedom
ML	Maximum likelihood
MMSE	Minimum mean square error
C-RAN	Cloud radio access network
IA	Interference alignment
RAN	Radio access network
BBU	Baseband unit
UE	User equipment
LAN	Local area network
DU	D2D user
DR	D2D receiver
SLNR	Signal-to-leakage noise ratio
QoS	Quality of services
SNR	Signal-to-noise ratio
PR	Primary receiver
PT	Primary transmitter
ST	Secondary transmitter
SR	Secondary receiver

SD	Spatial direction
SUS	Semi-orthogonal user selection
SU	Secondary user
SVD	Singular value decomposition
mmWave	Millimeter wave

Chapter 1
Introduction

1.1 5G Cellular Network

The fifth-generation (5G) cellular network is expected to support an ever increasing number of mobile devices with ubiquitous service access. It has been predicted that the traffic flow of mobile data will increase a thousand-fold by the year of 2020 [1]. It is widely agreed that compared to the 4G network, the 5G cellular network should achieve 1000 times the system capacity, 10 times the spectral efficiency, energy efficiency and data rate, and 25 times the average cell throughput [2]. To realize these ambitious goals, advanced 5G wireless technologies will be developed for higher spectrum efficiency, higher energy efficiency, and denser cell deployment. In this section, we introduce some of key features of 5G cellular system.

1.1.1 Multi-cell Cooperative Communications

Multi-cell joint processing at the base station (BS) side has been identified as a key approach for enhancing network performance. In stead of tuning their physical and link/MAC layer parameters (power level, time slot, subcarrier usage, beamforming coefficients, etc.) separately, or decoding independently of one another, the BSs coordinate their operations on the basis of global channel state and/or user data information that are exchanged over backhaul links among several cells [3, 4].

In the 5G cellular network, to support the extreme data traffic explosion, more aggressive spectrum reuse and much higher network capacity are needed. An effective way to increase the network capacity is to make the cells smaller [5]. On one hand, the deployment of small size cells shortens the distance between BSs, which facilitates the inter-cell cooperation. On the other hand, within small cells, the

© The Author(s) 2015
L. Yang, W. Zhang, *Interference Coordination for 5G Cellular Networks*,
SpringerBriefs in Electrical and Computer Engineering,
DOI 10.1007/978-3-319-24723-6_1

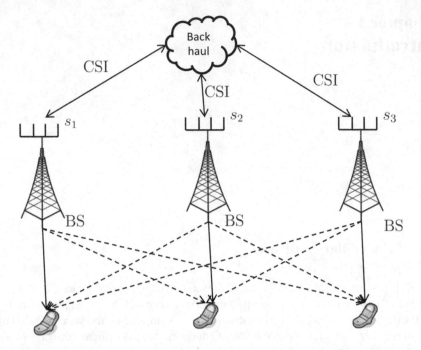

Fig. 1.1 The CSI-sharing BS cooperation, where the BSs acquire and exchange CSI but not data symbols

cell-edge users may receive strong interference from adjacent cells. Such inter-cell interference (ICI) must be coordinated jointly by BSs. Hence, multi-cell cooperation is not only suitable, but also necessary for 5G networks.

There are several possible levels of cooperations depending on the amount of overhead added on the backhaul and feedback channels.

Channel State Information (CSI)-Sharing Cooperation

In CSI sharing cooperation, the BSs share the CSI of both direct and interference links, which are obtained via over-the-air feedback channels, as shown in Fig. 1.1. The availability of global CSI allows BSs to coordinate their signaling strategies, such as power allocation, beamforming, and/or user scheduling. In this level of cooperation, the sharing of transmission data, or signal-level synchronization between the BSs is not necessary. Hence, the amount of overhead placed on backhaul is relatively modest.

It has been shown that the performance of cellular network can be improved even if the BSs only share the CSI [6–8].

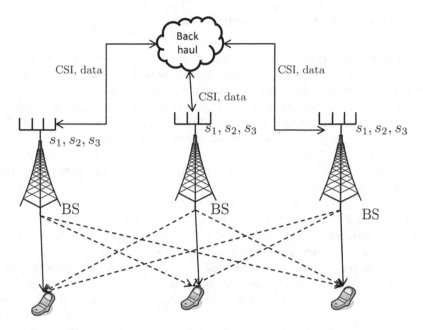

Fig. 1.2 The MIMO cooperation, where the BSs share both CSI and data symbols

MIMO Cooperation

When the BSs are linked with high-capacity, low-latency backhauls or connected to a high speed central processor, the cooperation between BSs can be more powerful. The BSs can share not only CSI, but also the full data signals of their respective users, as shown in Fig. 1.2. With this more powerful cooperation, a terminal is no longer served by an individual BS, but by a group of cells. The combined use of several antennas belonging to the BS in different cells to manage multiple data streams mimics the transmission over a MIMO channel, and hence is referred to as MIMO cooperation [3]. In some scenarios, the backhaul of the network is considered to be rate-limited. In that case, CSI is shared first, then only a sub-stream of user data or a quantized version of the signals are shared among the BSs.

The advantage of MIMO cooperation is that all propagation links (including interfering links) can be exploited to carry useful messages, and hence largely improve the performance of network [9–11].

Relay-Aided Cooperation

Instead of cooperating through backhaul links, it is also possible for the BSs to be assisted with separate relay nodes in each cell. It can be beneficial not only in strengthening the direct channel gain between BS and remote users, but also in dealing with inter-cell interference [12, 13].

1.1.2 Device-to-Device Communications

Unlike conventional cellular systems, where all communications must take place through BSs, device-to-device (D2D) communications allow two nearby users to establish direct communication links without traversing the BS or core network [14]. Potential D2D communications include peer-to-peer communication, relaying, multicasting, and machine-to-machine (M2M) communications. Figure 1.3 describes different types of D2D communications in cellular network.

As one of the key techniques in 5G cellular network, D2D communications can provide multiple performance benefits. First of all, it can largely improve the resource utilization and energy efficiency, while offloading cellular traffic and alleviating congestions [15], which are all the key demands for 5G system. Secondly, the D2D users can achieve high data rate and low latency attributing to the short distance of direct link. Third, D2D opens up new opportunities for proximity-based commercial services and peer-to-peer applications [16]. Moreover, D2D-enabled devices can also be a required feature in public safety networks [17].

In general, D2D communications can occur in cellular spectrum (i.e., in-band D2D communication) or in unlicensed spectrum (i.e., out-of-band D2D communication).

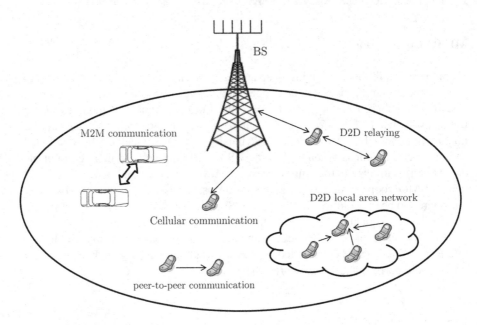

Fig. 1.3 D2D communications in cellular network

In-band D2D Communication

In in-band D2D communication, the cellular spectrum is used for both D2D and cellular links. The motivation for in-band communication is that the operators can have better control over cellular (licensed) spectrum. In general, the operators may have different levels of controls over D2D communications in aspects such as access authentication, connection maintenance, and resource allocation, etc. [18].

In-band D2D communication can be further divided into underlay and overlay categories. In underlay communications, the D2D links reuse the same spectrum resources of cellular users. On the other hand, in overlay communications, dedicated resources are allocated to D2D links.

Out-of-Band D2D Communication

In out-of-band D2D communications, the D2D links exploit unlicensed spectrum, which usually requires extra interface and other wireless techniques such as WiFi or bluetooth. The motivation for out-of-band communication is to eliminate the interference between D2D and cellular links.

1.1.3   Cognitive Radio Communications

Cognitive radio (CR) is an innovative software defined radio technique and is considered to be a promising technology to improve the utilization of the congested spectrum [19, 20]. Adopting CR in 5G network is motivated by the fact that in cellular network, some portions of spectrum that are allocated to a licensed user is often underutilized [2]. With CR techniques, secondary system can share spectrum bands with the licensed system, either on an interference-free basis or on an interference-tolerant basis.

In interference-free CR network, the unlicensed users must opportunistically reuse the underutilized spectrum which are defined as spectrum holes [21]. CR users may need to monitor and detect the unused spectrums via spectrum sensing. If there are multiple unlicensed system attempting to access the same spectrum, a coordination mechanism is also required to prevent the collision [22].

In interference-tolerate CR network, CR users can share the spectrum resource with a licensed system while keeping the interference under a threshold [23]. In comparison with interference-free CR networks, interference-tolerant CR networks can further improve the spectrum utilization and energy efficiency. However, the performance of CR systems may highly depend on the conditions of user density, interference threshold, and transmission behaviors of the licensed system [24, 25].

1.2   Interference Management: Challenges and Techniques

To accommodate high demand for heavy data traffic, the 5G cellular networks is featured by aggressive spectrum reuse, extreme densities of base stations and wireless devices, and integration of various communication systems. As a consequence, the management of interference is critical for maintaining the high performance of the network. In this section, some challenges of interference management for 5G network are presented first. Then, some useful interference coordination techniques are introduced.

1.2.1   Challenges of Interference Management

In the following, we introduce several types of interference that will appear in 5G networks, and discuss the challenges of managing them.

Other-Cell Interference

Cellular users, especially the ones at the cell edge, suffer two types of interference, which are self-cell interference from other users in the same cell and other-cell interference (OCI) from users in different cells. Methods of handling self-cell interference are by now well understood, i.e., the access of users in the same cell are generally orthogonal to each other through time-division multiple access (TDMA) or orthogonal frequency-division multiple access (OFDMA) [26]. However, due to the high density of cellular user equipments in 5G network, the orthogonality can not be guaranteed for the OCI. Hence, users in different cells may share the same time-frequency resource blocks, and cause interference to each other.

Cross-tier Interference

For in-band D2D communications, the D2D network coexists with cellular network in the same spectrum. In spit of those potential benefits, D2D techniques also pose many challenges and risks due to the difficulty of interference management, especially for underlay systems. In particular, the underlaid D2D signals become a new source of interference. Consequently, cellular links experience cross-tier interference from D2D transmissions while the D2D links suffer from not only the cross-tier interference from cellular transmissions but also the inter-D2D interference. Hence, an effective interference coordination mechanism is needed to ensure successful coexistence of cellular and D2D links.

Inter-system Interference

As mentioned before, for CR system to be integrated into 5G cellular network, the CR equipments (i.e., secondary system) must not affect the performance of cellular user equipments (i.e., primary system). If the CR systems are implemented in interference-free mode, the key is to successfully detect the available spectrum holes that spread out in the spectrum, geographic region, or spatial dimensions. If the CR systems are implemented in interference-tolerate mode, the major issue is how to reliably and practically control the mutual interference of CR and primary systems.

1.2.2 Interference Coordination Techniques

While it is challenging to effectively control different types of interference in the 5G network, there are some interference coordination techniques that show great potentials to overcome the challenging problems. Next, we introduce some promising techniques that can be used to deal with interference.

Beamforming Techniques

In the context of dealing with interference, beamforming refers to a class of signal processing techniques that can steer the transmitted signals such that their negative effect on receivers can be minimized. As a powerful tool of interference management, beamforming techniques have been used to improve the system performance in terms of a variety of measurements, such as signal-to-interference-and-noise ratio (SINR) [8], bit error rate (BER) [27], outage probability [28], and degrees of freedom (DoF) [29–31].

Power Control

Power control is an effective approach to reduce interference, which has been broadly utilized in wireless communication systems. In stead of allocating equal power to all users, unequal powers can be strategically distributed to different users based on certain parameters such as channel gain, user distance, or power consumption constraint, etc. As a result, not only the interference can be controlled, but also the energy efficiency can be improved [32]. Recently, there were considerable research efforts in power control techniques for D2D underlaid cellular networks [33, 34].

User Scheduling

In a wireless network, when there are more than one active users, it is possible to take advantage of multiuser diversity by scheduling transmissions to users with good channel conditions [35]. Generally speaking, multiuser diversity has provided another form of diversity in the system, which can be used to combat interference. Opportunistic user scheduling has been widely used in cellular networks and cognitive radio systems [36, 37].

Advanced Receiver Techniques

It is well known that interference-aware MIMO receivers can significantly suppress interference and largely improve the performance of the network. In particular, the Maximum Likelihood (ML) multiuser detection is known to minimize the bit error probability in a MIMO system [26]. Due to the complexity of implementing ML detection, a linear approximation of ML receiver can be used, which is referred to as MMSE receiver [38]. The advantage of MMSE receiver is that it well balances the noise enhancement and interference suppression. In addition, in order to handle strong interference that will be faced by cell-edge users in the next generation network, the receivers may need to have the capability to take advantage of the structure of the interference signals, including modulation constellation, coding scheme, channel, and resource allocation [39].

1.3   Structure of the Brief

5G cellular networks will embrace many advanced technologies and designs, which have shown great potentials of achieving high network capacity. However, it also brings new and high-risk challenges in dealing with the interference of the network. In order to achieve high capacity and high performance of 5G cellular network, it is essential to effectively coordinate various types of interference within the network. This Brief presents interference coordination techniques for 5G cellular networks. Along with a comprehensive overview of existing works, this Brief focuses on the design of practical interference coordination schemes that aim to deal with the inter-cell interference, cross-tier interference in D2D communications, and inter-system interference in cognitive radio networks.

In Chap. 2, we investigate the inter-cell interference coordination schemes for multi-cell MIMO network. Both CSI-sharing cooperation and MIMO cooperation for base stations are considered. With CSI-sharing cooperation, a spatial beam-forming scheme is proposed to eliminate the inter-cell interference and achieve optimal multiplexing gain. Then, the MIMO cooperation for base stations is also introduced in the form of cloud radio access network (C-RAN), which is an emerging network architecture designed for the 5G cellular network. The basic principle and implementation challenges of C-RAN is discussed.

In Chap. 3, we investigate the cross-tier interference coordination schemes for D2D communications and cellular networks. Several interference management schemes are proposed for different scenarios. We show that based on the proposed schemes, the interference generated on the cellular links is eliminated or well controlled, while the quality of service of D2D communications being guaranteed.

In Chap. 4, we study the inter-system interference coordination schemes for cognitive radio networks integrated in the cellular system. Both interference-free operation mode and interference-tolerate operation mode are considered. In interference-free mode, we introduce how to opportunistically exploit the spectrum holes of the cellular system. In interference-tolerate mode, the effect of user densities, interference threshold, and transmission behaviors of the systems are analyzed.

In Chap. 5, the conclusions are drawn and some future research directions are presented.

References

1. Wang, Y., Xu, J., & Jiang, L. (2015). Challenges of system-level simulations and performance evaluation for 5G wireless networks. *IEEE Access, 2*, 1553–1561.
2. Wang, C., Haider, F., Gao, X., You, X., Yang, Y., Yuan, D., Aggoune, H., Haas, H., Fletcher, S., & Hepsaydir, E. (2014). Cellular architecture and key technologies for 5G wireless communication networks. *IEEE Communications Magazine, 52*(2), 122–130.
3. Gesbert, D., Hanly, S., Huang, H., Shamai, S., Simeone, O., & Yu, W. (2010). Multi-cell MIMO cooperative networks: A new look at interference. *IEEE Journal on Selected Areas in Communications, 28*(9), 1380–1408.
4. Somekh, O., Simeone, O., Bar-Ness, Y., Haimovich, A., & Shamai, S. (2009). Cooperative multicell zero-forcing beamforming in cellular downlink channels. *IEEE Transactions on Information Theory, 55*(7), 3206–3219.
5. Andrews, J., Buzzi, S., Choi, W., Gesbert, D., Hanly, S., Lozano, A., Soong, A., & Zhang, J. (2014). What will 5G be. *IEEE Journal on Selected Areas in Communications, 32*(6), 1065–1082.
6. Shin, W., Lee, N., Lim, J., Shin, C., & Jang, K. (2011). On the design of interference alignment scheme for two-cell MIMO interference broadcast channels. *IEEE Transactions on Wireless Communications, 10*, 437–442.
7. Yang, L., & Zhang, W. (2013). Opportunistic interference alignment in heterogeneous two-cell uplink network. In *Proceedings of the IEEE International Conference on Communications (ICC 2013)*, Budapest, pp. 5448–5452, 9–13 June 2013
8. Shaverdian, A., & Nakhai, M. (2014). Robust distributed beamforming with interference coordination in downlink cellular networks. *IEEE Transactions on Communications, 62*(7), 2411–2421.
9. Foschini, G., Karakayali, K., & Valenzuela, R. (2006). Coordinating multiple antenna cellular networks to achieve enormous spectral efficiency. *The Proceeding of IEEE, 153*(4), 548–555.
10. Huh, H., Moon, S. H., Kim, Y. T., Lee, I., & Caire, G. (2011). Multi-cell MIMO downlink with cell cooperation and fair scheduling: A large system limit analysis. *IEEE Transactions on Information Theory, 57*(12), 7771–7786.

11. Hosseini, K., Yu, W., & Adve, R. (2014). Large-scale MIMO versus network MIMO for multicell interference mitigation. *IEEE Journal on Selected Topics in Signal Processing, 8*(5), 930–941.
12. Simeone, O., Somekh, O., Bar-Ness, Y., & Spagnolini, U. (2008). Throughput of low-power cellular systems with collaborative base stations and relaying. *IEEE Transactions on Information Theory, 54*(1), 459–467.
13. Zhai, C., Zhang, W., & Mao, G. (2012). Uncoordinated cooperative communications with spatially random relays. *IEEE Transactions on Wireless Communications, 11*(9), 3126–3135.
14. Asadi, A., Wang, Q., Mancuso, V. (2014). A survey on device-to-device communication in cellular networks. *IEEE Communications Surveys and Tutorials, 16*(4), 1801–1819. Fourth Quarter 2014.
15. Lin, X., Andrews, J., Ghosh, A., & Ratasuk, R. (2014). An overview of 3GPP device-to-device proximity services. *IEEE Communications Magazine, 52*(4), 40–48.
16. Fodor, G., Dahlman, E., Mildh, G., Parkvall, S., Reider, N., Miklos, G., & Turanyi, Z. (2012). Design aspects of network assisted device-to-device communications. *IEEE Communications Magazine, 50*(3), 170–177.
17. Doumi, T., Dolan, M., Tatesh, S., Casati, A., Tsirtsis, G., Anchan, K., & Flore, D. (2013). LTE for public safety networks. *IEEE Communications Magazine, 51*(2), 106–112.
18. Lei, L., Zhong, Z., & Shen, X. (2012). Operator controlled device-to-device communications in LTE-advanced networks. *IEEE Transactions on Wireless Communications, 19*(3), 96–104.
19. Mitola, J. (2000). Cognitive radio: An integrated agent architecture for software defined radio. Ph.D. dissertation, KTH, Stockholm.
20. Wang, Z., & Zhang, W. (2015). *Opportunistic spectrum sharing in cognitive radio networks.* Springer. ISBN:978-3-319-15541-8.
21. Kolodzy, P., et al. (2001). Next generation communications: Kickoff meeting. In *Proceeding of the DARPA*, Arlington. 17 Oct 2001.
22. Zhai, C., Zhang, W., & Mao, G. (2014). Cooperative spectrum sharing between cellular and ad-hoc networks. *IEEE Transactions on Wireless Communications, 13*(7), 4025–4037.
23. Hamdi, K., Zhang, W., & Letaief, K. B. (2009). opportunistic spectrum sharing in cognitive MIMO wireless networks. *IEEE Transactions on Wireless Communications, 8*(8), 4098–4109.
24. Perlaza, S. M., Fawaz, N., Lasaulce, S., & Debbah, M. (2010). From spectrum pooling to space pooling: Opportunistic interference alignment in MIMO cognitive networks. *IEEE Transactions on Signal Processing, 58*(7), 3728–3741.
25. Yang, L., Zhang, W., Zheng, N., & Ching, P. C. (2014). Opportunistic user scheduling in MIMO cognitive radio networks. In *Proceedings of the IEEE International Conference on Acoustics, Speech and Signal Processing (ICASSP 2014)*, Florence, pp. 7303–7307, 4–9 May 2014.
26. Andrews, J., Choi, W., & Heath, R., Jr. (2007). Overcoming interference in spatial multiplexing MIMO cellular networks. *IEEE Transactions on Wireless Communications, 14*(6), 95–104.
27. Chae, C., Mazzarese, D., Jindal, N., & Heath, R., Jr. (2008). Coordinated beamforming with limited feedback in the MIMO broadcast channel. *IEEE Journal on Selected Areas in Communications, 26*(8), 1505–1515.
28. Yang, L., Zhang, W., & Jin, S. (2015). Interferene alignment in device-to-device LAN underlaying cellular network. *IEEE Transactions on Wireless Communications, 14*(7), 3715–3723.
29. Cadambe, V., & Jafar, S. (2008). Interference alignment and degrees of freedom of the K-user interference channel. *IEEE Transactions on Information Theory, 54*(8), 3425–3441.
30. Yang, L., & Zhang, W. (2014). Interference alignment with asymmetric complex signaling on MIMO X channels. *IEEE Transactions on Communications, 62*(10), 3560–3570.
31. Yang, L., & Zhang, W. (2015). On degrees of freedom region of three-user MIMO interference channels. *IEEE Transactions on Signal Processing, 63*(3), 590–603.
32. Andrews, J., & Meng, T. (2003). Optimum power control for successive interference cancellation with imperfect channel estimation. *IEEE Transactions on Wireless Communications, 2*(2), 375–383.

33. Yu, C.-H., Doppler, K., Ribeiro, C. B., & Tirkkonen, O. (2011). Resource sharing optimization for device-to-device communication underlaying cellular networks. *IEEE Transactions on Wireless Communications, 10*(8), 2752–2763.
34. Lee, N., Lin, X., Andrews, J., & Heath, R., Jr. (2015). Power control for D2D underlaid cellular networks: Modeling, algorithms, and analysis. *IEEE Journal on Selected Areas in Communications, 33*(1), 1–13.
35. Viswanath, P., Tse, D., & Laroia, R. (2002). Opportunistic beamforming using dumb antennas. *IEEE Transactions on Information Theory, 48*(6), 1277–1294.
36. Hamdi, K., Zhang, W., & Letaief, K. B. (2009). Opportunistic spectrum sharing in cognitive MIMO wireless networks. *IEEE Transactions on Wireless Communications, 8*(8), 4098–4109
37. Yoo, T., & Goldsmith, A. (2006). On the optimality of multiantenna broadcast scheduling using zero-forcing beamforming. *IEEE Journal on Selected Areas in Communications, 24*(3), 528–541.
38. Choi, W., Andrews, J., & Heath, R., Jr. (2007). Multiuser antenna partitioning for MIMO-CDMA system. *IEEE Transactions on Vehicular Technology, 56*(5), 2448–2456.
39. Nam, W., Bai, D., Lee, J., & Kang, I. (2014). Advanced interference managment for 5G cellular network. *IEEE Communications Magazine, 52*(5), 52–60.

Chapter 2
Interference Coordination in Multi-cell MIMO Networks

2.1 Introduction

The 5G cellular network is featured by the dense deployment of BSs with small cell range and extensive reuse of spectrum resources. However, these features may have huge impact on the performance of cell-edge users. The extensive reuse of spectrum may cause severe interference because multiple users in the boundaries of multiple cells are allocated to the same resource block. Since the neighboring BSs are also located nearby, the users are exposed to strong inter-cell interference as well. In this case, if the BSs operate without cooperation, the network will become interference-limited, i.e., the signal-to-interference-and-noise ratio (SINR) at users cannot be improved by simply increasing the transmit power, as higher transmit power also creates stronger interference. In this chapter, we investigate interference coordination techniques with BS cooperation to combat the inter-cell interference for cell-edge users. As mentioned in Chap. 1, the BSs can cooperate with each other either by sharing the CSI or by sharing both CSI and users' data, which are referred to as CSI-sharing BS cooperation and MIMO BS cooperation, respectively. In addition, the requirement for CSI-sharing cooperation is much less than MIMO cooperation.

While both cooperation mechanisms will be addressed in this chapter, our main focus is on interference coordination techniques based on CSI-sharing cooperation. In particular, a practical 3-cell MIMO downlink network is considered. As shown in Fig. 2.1, three users sharing the same spectrum resources are in the cell-intersection area, each of which receives signals from one BS and interference

© The Author(s) 2015
L. Yang, W. Zhang, *Interference Coordination for 5G Cellular Networks*,
SpringerBriefs in Electrical and Computer Engineering,
DOI 10.1007/978-3-319-24723-6_2

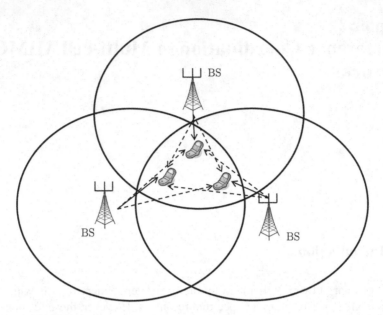

Fig. 2.1 Three-cell MIMO network

from other two BSs. With CSI-sharing cooperation, a beamforming scheme is proposed to coordinate the interference among the users. The performance of the proposed scheme is evaluated in terms of degrees of freedom (DoF), which provides effective measurement on the interference control level. It is shown that the optimal DoF region of the network can be achieved. The details of the scheme, including background, model, related works, key techniques and designs, will be presented in Sect. 2.2.

Next, we discuss the base station cooperation on the level of sharing both CSI and the data of users, which has the potential to transform the network into a large scale virtual MIMO system. Based on such cooperation, the interference link will not only be mitigated, but also be exploited to carry useful messages, which can largely enhance the network capacity. However, in spite of the obvious advantages, it is difficult to be implemented in practice due to its high demands on backhaul capacity and strict requirements on signal synchronization and channel feedback. A recently proposed network architecture, called cloud radio access network (C-RAN) [1], is thought of as a possible solution for the practical implementation of such "network" MIMO coordinated transmission concept. In Sect. 2.3, the general idea and some design challenges of C-RAN will be discussed.

The rest of the chapter is organized as follows. In Sect. 2.2, the interference coordination for a three-cell MIMO network is investigated with CSI-sharing BS cooperation. In Sect. 2.3, the C-RAN architecture based on virtual MIMO BS cooperation is introduced. Section 2.4 summarizes the Chapter.

2.2 Three-Cell MIMO Network

In this Section, we investigate interference coordination techniques for cellular networks under CSI-sharing BS cooperation.

First we introduce "degrees of freedom" (DoF) that is used to measure the effectiveness of interference management schemes. The definition of DoF is given as [2]

$$d \triangleq \lim_{\rho \to \infty} \frac{C(\rho)}{\log_2(\rho)} \tag{2.1}$$

where $C(\rho)$ is the sum capacity of the network with SNR ρ.

From (2.1) we can see that DoF provides accurate capacity approximation in high signal-to-noise ratio (SNR) region. This implies that with high SNR, achieving high DoF is equivalent to achieving high capacity of the network or high rate of the user. Note that with small cell arrangement, the distances between BS and user equipments are shortened, which means it is possible to ensure high SNR in the network. In addition, DoF in a wireless network can also be seen as the measurement of the number of interference-free spatial signaling dimensions that are accessible in the network [3].

Due to its importance in characterizing wireless network capacity, the optimal achievable DoF has been extensively investigated for a variety of wireless networks under different conditions. Next, we briefly review some related works that focus on DoF characterization of cellular network. The works in [4–6] focus on two-cell scenarios. In [4], an interference alignment scheme is designed for two-cell interfering broadcast networks where each cell has two users at the intersection of cells. It is shown in [4] that the optimal DoF can be achieved. The same system model is also studied in [5], where the optimal DoF is achieved with feedbacks only within a cell. The case of multiple users is considered in [6], where opportunistic beamforming scheme is proposed to exploit the benefit of multiuser diversity. When the number of cells is more than two, the difficulty of achieving optimal DoF is largely increased because each user suffers from more than one interference signals. As mentioned in Chap. 1, signals of users in the same cell can be made orthogonal to each other in time and/or frequency domain. Under this premise, it can be expected that in the intersection of K multiple cells, there would be at most K users using the same spectrum resources and each of them belongs to a different cell. Hence, the network mimics a K-user interference channels, where each user receives desired signals from one transmitter and interference from other transmitters. The theoretical DoF of K-user interference channels was studied in [3, 7, 8, 13, 22], where each BS and each user are equipped with M_T and M_R antennas, respectively. It was shown in [7] that the optimal DoF of $\frac{K}{2}$ can be achieved asymptotically when $M_T = M_R = 1$. For multiple-antenna cases, it was shown in [3] that if $\eta = \frac{\max(M_T, M_R)}{\min(M_T, M_R)}$ is an integer, each user can achieve DoF of $\min(M_T, M_R)\frac{\eta}{\eta+1}$ when $K > \eta$. The result of [3], established originally over time-varying channels, was extended to

constant channels in [8, 22]. More general results are obtained in [13] for three-user
case only. According to [13], the outer-bound DoF of each link of 3-user interference
channels equals DoF^*, where

$$DoF^* = \min\{\frac{\kappa}{2\kappa-1}M , \frac{\kappa}{2\kappa+1}N\} \tag{2.2}$$

with $N = \max\{M_T , M_R\}$, $M = \min\{M_T , M_R\}$ and $\kappa = \lceil \frac{M}{N-M} \rceil$.

Note that in theory, the DoF can be either integers or fractions. However, in
practical scenarios like cellular networks, the achievable DoF is usually integers.
Hence, it is also of great interest to investigate the optimal integer DoF of the
network [9–14]. The most recent results for three-user case in [13, 14] show that
the optimal integer DoF for each user is $\lfloor DoF^* \rfloor$, assuming that each user always
has the same DoF.

The focus of this Section is three-cell MIMO network, which is equivalent
to three-user interference channels. We investigate some unsettled issues on the
achievable DoF of three-user interference channels. Note that in previous works
[13, 14], each user must have the same DoF. As a result, the DoF characterization
of each user is the same as the characterization of the sum DoF of the network.
Nevertheless, since the users in cellular networks may have different number of
data streams and multiplexing gain, it is necessary to consider the cases when the
users are allowed to have different DoF. Hence, instead of only studying the *sum
DoF* of the network, the *DoF region* should be considered. Let d_i denote the DoF
of user i and $D = d_1 + d_2 + d_3$. It was shown in [13] that the outer bound of *sum
DoF* of 3-user interference channel is $D \leq 3DoF^*$. In addition, if the DoF of each
user denotes one coordinate, the DoF region of 3-user interference channel would
be a 3-dimensional space that is closed by multiple planes. As we can see, the
sum DoF $D = 3DoF^*$ can be seen as one plane, and the space under the plane,
$D \leq 3DoF^*$, can certainly be seen as an outer bound of DoF region. However, this
region is very loose. In the following, we first derive an outer bound DoF region.
Then, a spatial beamforming scheme is proposed that can achieve all the integer
DoF inside the DoF region. Hence, the derived DoF region is the exact optimal
achievable integer DoF region of three-user interference channels. In the following,
the details of the scheme, including system model, main results, key techniques and
designs, are presented.

2.2.1 Interference Alignment

In this section, we introduce a novel technique that has been extensively studied in
recent years, which is referred to as interference alignment. The emergence of IA
brings us new insights on interference management and is very powerful in DoF
characterization. In this section, we first introduce the basic concept of IA. Then,
we review an advanced IA technique, called "subspace alignment chain", which is
proposed in [13].

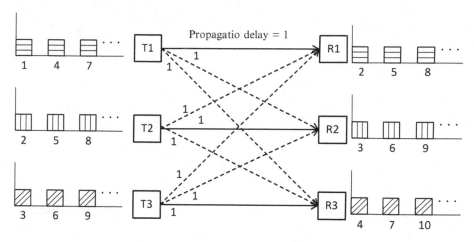

Fig. 2.2 TDMA transmission scheme

Basic Concept of Interference Alignment

We first use an example in [7] to explain the basic concept of IA. Three single-antenna transmitters, T_1, T_2, and T_3, communicate with three single-antenna receivers, R_1, R_2, and R_3, respectively. The signal from one transmitter constitutes interference on two unintended receivers. Conventionally, orthogonal transmission scheme such as TDMA [15] can be used for medium access, i.e., each transmitter-receiver pair transmits in one third of entire period of time while other two pairs remain silent. As a result, each pair can have one third of the transmission time. The TDMA transmission is shown in Fig. 2.2.

Now, it is interesting to know if we can do better than that. In [7], it is assumed that the propagation delay can be controlled so that the delay from one transmitter to the intended receiver is one symbol period, whereas to its unintended receivers is two symbol period. Then, each transmitter only transmits at odd time slots and remains silent at even time slots. Consequently, at each receiver, the desired signal arrives at even time slots while all interference signals are received only at odd time slots, as shown in Fig. 2.3.

In that case, each transmitter-receiver pair can have half of the degrees of freedom, which is more efficient than TDMA. At each receiver, the two interference signals are aligned at the same time slots, which is referred to as interference alignment. As we can see, the above artificial delay model is not realistic. However, the IA is not limited in time. It can be also implemented in space, frequency, or even signal domains. According to [16], IA is defined as a construction of signals in a way that they cast overlapping shadows at the receivers where they constitute interference, while remaining distinguishable at the receivers where they are desired.

In general, IA schemes can be broadly classified into two categories, which are signal vector space IA [2, 7, 9–14, 16–19] and signal level IA [20–23]. Specifically, in signal vector space IA, by exploiting the distinct linear transformation (channel

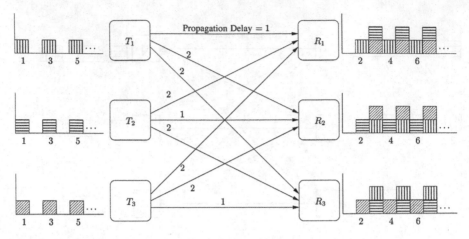

Fig. 2.3 Interference alignment transmission scheme [7]

matrix) between each transmitter-receiver pair, the transmitters perform linear precoding to rotate the signal vectors on each link. This type of IA inherits the advantages of linear processing. It is tractable, linearly decodable, and applicable for any channels [17]. On the other hand, signal-level IA uses structured coding to align interference in the signal-level space. The advantage of signal-level IA is that it does not need the distinct rotation of channel matrices, which may not always be available. Since signal-level IA is derived from the deterministic channel model in [24], they are most suitable for channels with real coefficients. Moreover, the decoding methods for signal-level IA are more complicated due to the non-linear effect. Hence, considering the channel conditions and the complexity of implementations in practice, vector space IA is adopted in our proposed scheme for cellular networks.

In addition to cellular network, IA techniques have been applied to many other networks such as cognitive radio networks [25–27], multi-hop networks [28–31], device-to-device network [32], and storage system [33], etc.

Subspace Alignment Chain [13]

Since the origins of IA [7, 19], the technique has been widely investigated and improved in many ways. Next, we review a recently developed technique, called "subspace alignment chain" [13], which will be used in our proposed scheme.

Two ends of an alignment chain correspond to the signals that are nulled at one unintended receiver and cause interference at the other. For the signals in between, each of which is aligned with another interference signal at each of those undesired receivers. The length of each chain is defined as the number of signals that participate in it.

Let $\mathbf{V}^j_{i(n)} \in \mathbb{C}^{N \times Q_j}$ denote the sth Q_j-dimensional subspace transmitted by transmitter i which participates in the chain that originates from transmitter j. Consider one alignment chain originating from transmitter 1, where $\mathbf{V}^1_{1(1)}$ is nulled at receiver 2 but causes an interference dimension at receiver 3. The second signal, $\mathbf{V}^1_{2(1)}$ from transmitter 2, should be aligned with $\mathbf{V}^1_{1(1)}$ on receiver 3, so that no more interference dimension is generated on receiver 3. Then, if $\mathbf{V}^1_{2(1)}$ can be nulled at receiver 1, the chain is finished. Otherwise, transmitter 3 should send a vector, $\mathbf{V}^1_{3(1)}$, which is aligned with $\mathbf{V}^1_{2(1)}$ on receiver 1. The chain will keep going until the zero-forcing can be achieved. Mathematically, it can be expressed as follows,

$$
\underbrace{\begin{bmatrix}
\mathbf{H}_{21} & 0 & \cdots & \cdots & 0 \\
\mathbf{H}_{31} & -\mathbf{H}_{32} & 0 & \cdots & 0 \\
0 & \mathbf{H}_{12} & -\mathbf{H}_{13} & 0 & \cdots \\
\vdots & \ddots & & \cdots & \\
0 & \cdots & \cdots & \mathbf{H}_{ri} & -\mathbf{H}_{rj} \\
0 & 0 & \cdots & 0 & \mathbf{H}_{ij}
\end{bmatrix}}_{\mathbf{H} \in \mathbb{C}^{M(S+1) \times S \cdot N}}
\underbrace{\begin{bmatrix}
\mathbf{V}^1_{1(1)} \\
\mathbf{V}^1_{2(1)} \\
\mathbf{V}^1_{3(1)} \\
\mathbf{V}^1_{1(2)} \\
\vdots \\
\mathbf{V}^1_{i(n)} \\
\mathbf{V}^1_{j(n)}
\end{bmatrix}}_{\mathbf{V} \in \mathbb{C}^{S \cdot N \times Q_1}} = \mathbf{0} \qquad (2.3)
$$

where S is the length of the chain, which equals the number of subspaces, $\mathbf{V}^1_{i(n)}$, that participate in the chain.

As can be seen, zero-forcing can be performed at the end of chain when the matrix $\mathbf{H}$ turns into a "fat" matrix, i.e., $S \cdot N > (S+1)M \Rightarrow S > \frac{M}{N-M}$. Hence, the length of the shortest chain can be expressed as

$$
S = \begin{cases} \lceil \frac{M}{N-M} \rceil + 1 & \text{when } \frac{M}{N} = \frac{p}{p+1} \\ \lceil \frac{M}{N-M} \rceil & \text{when } \frac{M}{N} \neq \frac{p}{p+1} \end{cases} \qquad (2.4)
$$

where $p = 1, 2, 3, \cdots, +\infty$. Note that for any chain that is longer than S, $\mathbf{H}$ will always be a "fat" matrix, which means for each antenna configuration, there exist multiple chains with length equal to S, $S+1, \cdots$.

The chains with the length of S are referred to as the *original* alignment chains. It can be seen there are three *original* chains. Each chain originates from one transmitter, i.e., $j = 1, 2, 3$. The three *original* chains can be expressed as follows

$$
\mathbf{0} \xleftrightarrow{R_2} \mathbf{V}^1_{1(1)} \xleftrightarrow{R_3} \mathbf{V}^1_{2(1)} \xleftrightarrow{R_1} \mathbf{V}^1_{3(1)} \xleftrightarrow{R_2} \mathbf{V}^1_{1(2)} \cdots \mathbf{0}
$$

$$
\mathbf{0} \xleftrightarrow{R_3} \mathbf{V}^2_{2(1)} \xleftrightarrow{R_1} \mathbf{V}^2_{3(1)} \xleftrightarrow{R_2} \mathbf{V}^2_{1(1)} \xleftrightarrow{R_3} \mathbf{V}^2_{2(2)} \cdots \mathbf{0}
$$

$$
\mathbf{0} \xleftrightarrow{R_1} \mathbf{V}^3_{3(1)} \xleftrightarrow{R_2} \mathbf{V}^3_{1(1)} \xleftrightarrow{R_3} \mathbf{V}^3_{2(1)} \xleftrightarrow{R_1} \mathbf{V}^3_{3(2)} \cdots \mathbf{0} \qquad (2.5)
$$

where $\mathbf{V}^1_{1(1)} \overset{R_3}{\longleftrightarrow} \mathbf{V}^1_{2(1)}$ means that the interference generated by transmitter 1 $\mathbf{V}^1_{1(1)}$ and the one generated by transmitter 2 $\mathbf{V}^1_{2(1)}$ are aligned at receiver 3, i.e., $\mathbf{H}_{31}\mathbf{V}^1_{1(1)} = \mathbf{H}_{32}\mathbf{V}^1_{2(1)}$, as shown in (2.3).

2.2.2 System Model and Main Result

Let M_T and M_R denote the number of antennas on each BS and user, respectively. We consider a fully connected three-user MIMO interference channel with M_T and M_R antennas at each transmitter and each receiver, respectively. As shown in Fig. 2.4, transmitter i sends messages to receiver i ($i = 1, 2, 3$) and causes interference to other two receivers. Let $N = \max\{M_R, M_T\}$ and $M = \min\{M_R, M_T\}$. Note that when $\frac{M}{N} \leq \frac{1}{2}$, the DoF region is just the combination of single user and/or cooperation DoF outer bounds [2, 13], which can be trivially achieved. Hence, in this Chapter, we focus on the region $\frac{1}{2} < \frac{M}{N} < 1$, where the DoF region remains an open problem.

Let $\mathbf{H}_{ji} \in \mathbb{C}^{M_R \times M_T}$ denote the channel matrix between transmitter i and receiver j. We assume that all channel matrices are sampled from continuous complex Gaussian distributions and each entry of $\mathbf{H}_{ij}$ is independent and identically distributed (i.i.d.). Based on CSI-sharing BS cooperation, the global CSI is available at all transmitters.

The received signals on receiver j can be expressed as

$$\mathbf{y}_j = \sum_{i=1}^{3} \mathbf{H}_{ji}\mathbf{B}_i\mathbf{m}_i + \mathbf{z}_j \qquad (2.6)$$

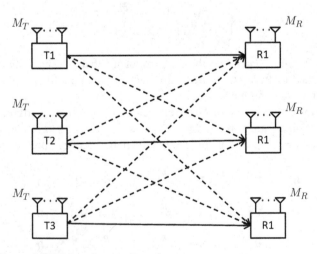

Fig. 2.4 Three-user MIMO interference channel

where $\mathbf{y}_j \in \mathbb{C}^{M_R \times 1}$ denotes the received signal, $\mathbf{B}_i \in \mathbb{C}^{M_T \times d_i}$ denotes the beamforming matrix of transmitter i, $\mathbf{m}_i \in \mathbb{C}^{d_i \times 1}$ denotes the original message vector from transmitter i, and $\mathbf{z}_j \in \mathbb{C}^{M_R \times 1}$ denotes the white Gaussian noise at receiver j.

In addition, in the case of $M_T > M_R$ (i.e., $N = M_T$ and $M = M_R$), each channel matrix has a $(N - M)$-dimensional null space. Let nullspace$\{\mathbf{H}_{ij}\}$ denote the span of the null space of $\mathbf{H}_{ij}$. The following conditions are satisfied almost surely because the channels are generic and $N \leq 2M$.

$$\text{nullspace}\{\mathbf{H}_{21}\} \cap \text{nullspace}\{\mathbf{H}_{31}\} = \varnothing \tag{2.7}$$

$$\text{nullspace}\{\mathbf{H}_{12}\} \cap \text{nullspace}\{\mathbf{H}_{32}\} = \varnothing \tag{2.8}$$

$$\text{nullspace}\{\mathbf{H}_{13}\} \cap \text{nullspace}\{\mathbf{H}_{23}\} = \varnothing \tag{2.9}$$

Let ρ denote the power constraints on each transmitter and $\mathbf{R}_i(\rho)$ denote the achievable rate of user i. The DoF of user i is defined as $\lim_{\rho \to \infty} \frac{\mathbf{R}_i(\rho)}{\log(\rho)}$, which can be interpreted as the number of independent signaling dimensions or streams available for user i. Further, note that d_i is the number of signals sent by transmitter i. If the desired signals on each receiver can be linearly decoded, d_i would be equal to the DoF of link i.

The main results are given in the following two theorems.

Theorem 2.1. *In 3-user interference channels where each transmitter is equipped with M_T antennas and each receiver is equipped with M_R antennas, the outer bound of DoF region, $\mathbf{R}(d_1, d_2, d_3)$, is*

$$\begin{cases} 2td_i + 2td_j + (2t-1)d_k \leq (3t-1)N \\ (2t-1)d_i + 2(t-1)d_j + 2(t-1)d_k \leq (3t-2)M \\ (2t-1)d_i + (2t-1)d_j + (2t-1)d_k \leq (3t-1)M \\ d_i + d_j \leq N \end{cases} \tag{2.10}$$

for $\frac{M}{N} \in [\frac{3t-2}{3t-1}, \frac{3t-1}{3t})$, $(t = 1, 2, \cdots \infty)$

$$\begin{cases} 2td_i + 2td_j + (2t-1)d_k \leq 3tM \\ (2t+1)d_i + 2td_j + 2td_k \leq 3tN \\ d_i + d_j \leq N \end{cases} \tag{2.11}$$

for $\frac{M}{N} \in [\frac{3t-1}{3t}, \frac{3t}{3t+1})$, and

$$\begin{cases} (2t+1)d_i + 2td_j + 2td_k \leq (3t+1)M \\ (2t+1)d_i + (2t+1)d_j + (2t+1)d_k \leq (3t+1)N \\ d_i + d_j \leq N \end{cases} \tag{2.12}$$

for $\frac{M}{N} \in [\frac{3t}{3t+1}, \frac{3t+1}{3t+2})$, *where* $N = \max\{M_T, M_R\}$, $M = \min\{M_T, M_R\}$, i, j, $k =$ 1, 2, 3 *and* $i \neq j \neq k$.

The proof of **Theorem 2.1** is in [34].

Theorem 2.2. *For the 3-user MIMO interference channels with M_T and M_R antennas on each transmitter and receiver, respectively and $\frac{1}{2} \leq \frac{\min\{M_T, M_R\}}{\max\{M_T, M_R\}} < 1$, the DoF of user i, d_i (where $i = 1,2,3$ and d_i is integer), can be achieved if and only if (2.10), (2.11) and (2.12) are satisfied.*

The converse proof follows directly from **Theorem 2.1**, i.e., any DoF that does not satisfy (2.10), (2.11) and (2.12) cannot be achieved for sure as (2.10), (2.11) and (2.12) is outer bound. The achievability proof is given in Sect. 2.2.3 by proposing a linear beamforming scheme. Details of the proof of **Theorem 2.2** is in [34].

Remark 2.1. In three-cell cellular networks, the DoF tuple of (d_1, d_2, d_3) can be seen as the numbers of data streams of each user that can be transmitted free from interference, as long as (2.10), (2.11) and (2.12) is satisfied.

2.2.3 A Beamforming Scheme

In this section, we present a beamforming scheme for the three-user interference channels. Specifically, we present the design of beamforming matrix of each BS in the aim of achieving optimal DoF region of the network.

In [13], the *original* chain (2.5) is designed to achieve the point with coordinates $(\lfloor DoF^* \rfloor, \lfloor DoF^* \rfloor, \lfloor DoF^* \rfloor)$ in the space of DoF region. However, there are some points in the region that cannot be achieved with the scheme in [13]. This is because those points have unequal values of d_1, d_2 and d_3, which means the signal spaces on the users are unbalanced, and *original* chain itself does not have enough flexibility to distribute the signal space accordingly. Next, we propose a beamforming scheme design based on three types of chains: *original* chains, long chains (with length $\bar{S} = S + 1$), and null space of interference channels. Note that in theory, the length of each chain can be arbitrarily long. Hence, it is not trivial to select the chains that only with length of S and $S + 1$.

The beamforming matrix of transmitter i is designed as

$$\mathbf{B}_i = \begin{bmatrix} \mathbf{V}_i & \bar{\mathbf{V}}_i & \mathbf{U}_i \end{bmatrix} \tag{2.13}$$

where $\mathbf{V}_i$ is composed of all the subspaces from transmitter i that participate in the *original* chains (as shown in (2.5)), i.e.,

$$\mathbf{V}_i = \begin{bmatrix} \mathbf{V}^1_{i(1)} & \mathbf{V}^1_{i(2)} & \cdots & \mathbf{V}^2_{i(1)} & \mathbf{V}^2_{i(2)} & \cdots & \mathbf{V}^3_{i(1)} & \mathbf{V}^3_{i(2)} & \cdots \end{bmatrix} \tag{2.14}$$

and $\bar{\mathbf{V}}_i$ is composed of all the subspaces from transmitter i that participate in long chains (which is similar to (2.5) but with one more subspace at the end of each chain), i.e.,

$$\bar{\mathbf{V}}_i = \left[\bar{\mathbf{V}}^1_{i(1)} \ \bar{\mathbf{V}}^1_{i(2)} \ \cdots \ \bar{\mathbf{V}}^2_{i(1)} \ \bar{\mathbf{V}}^2_{i(2)} \ \cdots \ \bar{\mathbf{V}}^3_{i(1)} \ \bar{\mathbf{V}}^3_{i(2)} \ \cdots \right] \tag{2.15}$$

where $\bar{\mathbf{V}}^j_{i(\bar{n})} \in \mathbb{C}^{N \times \bar{Q}_j}$.

In addition, $\mathbf{U}_i = \left[\mathbf{U}^1_i \ \mathbf{U}^2_i \right] \in \mathbb{C}^{M \times q_i}$, which is designed as follows

$$\mathbf{H}_{21}\mathbf{U}^1_1 = \mathbf{0}, \ \mathbf{H}_{31}\mathbf{U}^2_1 = \mathbf{0}$$
$$\mathbf{H}_{12}\mathbf{U}^1_2 = \mathbf{0}, \ \mathbf{H}_{32}\mathbf{U}^2_2 = \mathbf{0}$$
$$\mathbf{H}_{13}\mathbf{U}^1_3 = \mathbf{0}, \ \mathbf{H}_{23}\mathbf{U}^2_3 = \mathbf{0} \tag{2.16}$$

Each part of $\mathbf{U}_i$ is nulled at one of the unintended receivers and causes interference to the other. Let q_{ji} denote the number of interference dimensions generated on receiver j by $\mathbf{U}_i$. We have $\mathbf{U}^1_1 \in \mathbb{C}^{N \times q_{31}}$, $\mathbf{U}^2_1 \in \mathbb{C}^{N \times q_{21}}$, $\mathbf{U}^1_2 \in \mathbb{C}^{N \times q_{32}}$, $\mathbf{U}^2_2 \in \mathbb{C}^{N \times q_{12}}$, $\mathbf{U}^1_3 \in \mathbb{C}^{N \times q_{23}}$, and $\mathbf{U}^2_3 \in \mathbb{C}^{N \times q_{13}}$.

In summary, all the beamforming subspaces of three transmitters can be designed as follows,

$$\mathbf{0} \xleftrightarrow{R_2} \mathbf{V}^1_{1(1)} \xleftrightarrow{R_3} \mathbf{V}^1_{2(1)} \xleftrightarrow{R_1} \mathbf{V}^1_{3(1)} \xleftrightarrow{R_2} \mathbf{V}^1_{1(2)} \cdots \mathbf{V}^1_{i(\lceil \frac{S}{3} \rceil)} \xleftrightarrow{R_k} \mathbf{0} \qquad (a)$$

$$\mathbf{0} \xleftrightarrow{R_3} \mathbf{V}^2_{2(1)} \xleftrightarrow{R_1} \mathbf{V}^2_{3(1)} \xleftrightarrow{R_2} \mathbf{V}^2_{1(1)} \xleftrightarrow{R_3} \mathbf{V}^2_{2(2)} \cdots \mathbf{V}^1_{j(\lceil \frac{S}{3} \rceil)} \xleftrightarrow{R_i} \mathbf{0} \qquad (b)$$

$$\mathbf{0} \xleftrightarrow{R_1} \mathbf{V}^3_{3(1)} \xleftrightarrow{R_2} \mathbf{V}^3_{1(1)} \xleftrightarrow{R_3} \mathbf{V}^3_{2(1)} \xleftrightarrow{R_1} \mathbf{V}^3_{3(2)} \cdots \mathbf{V}^1_{k(\lceil \frac{S}{3} \rceil)} \xleftrightarrow{R_j} \mathbf{0} \qquad (c)$$

$$\mathbf{0} \xleftrightarrow{R_2} \bar{\mathbf{V}}^1_{1(1)} \xleftrightarrow{R_3} \bar{\mathbf{V}}^1_{2(1)} \xleftrightarrow{R_1} \bar{\mathbf{V}}^1_{3(1)} \xleftrightarrow{R_2} \bar{\mathbf{V}}^1_{1(2)} \cdots \bar{\mathbf{V}}^1_{i(\lceil \frac{S}{3} \rceil)} \xleftrightarrow{R_k} \bar{\mathbf{V}}^1_{j(\lceil \frac{S+1}{3} \rceil)} \xleftrightarrow{R_i} \mathbf{0} \quad (d)$$

$$\mathbf{0} \xleftrightarrow{R_3} \bar{\mathbf{V}}^2_{2(1)} \xleftrightarrow{R_1} \bar{\mathbf{V}}^2_{3(1)} \xleftrightarrow{R_2} \bar{\mathbf{V}}^2_{1(1)} \xleftrightarrow{R_3} \bar{\mathbf{V}}^2_{2(2)} \cdots \bar{\mathbf{V}}^1_{j(\lceil \frac{S}{3} \rceil)} \xleftrightarrow{R_i} \bar{\mathbf{V}}^2_{k(\lceil \frac{S+1}{3} \rceil)} \xleftrightarrow{R_j} \mathbf{0} \quad (e)$$

$$\mathbf{0} \xleftrightarrow{R_1} \bar{\mathbf{V}}^3_{3(1)} \xleftrightarrow{R_2} \bar{\mathbf{V}}^3_{1(1)} \xleftrightarrow{R_3} \bar{\mathbf{V}}^3_{2(1)} \xleftrightarrow{R_1} \bar{\mathbf{V}}^3_{3(2)} \cdots \bar{\mathbf{V}}^1_{k(\lceil \frac{S}{3} \rceil)} \xleftrightarrow{R_j} \bar{\mathbf{V}}^3_{i(\lceil \frac{S+1}{3} \rceil)} \xleftrightarrow{R_k} \mathbf{0} \quad (f)$$

$$\mathbf{0} \xleftrightarrow{R_2} \mathbf{U}^1_1, \ \mathbf{0} \xleftrightarrow{R_3} \mathbf{U}^2_2, \ \mathbf{0} \xleftrightarrow{R_1} \mathbf{U}^1_3; \mathbf{U}^2_1 \xleftrightarrow{R_3} \mathbf{0}, \ \mathbf{U}^1_2 \xleftrightarrow{R_1} \mathbf{0}, \ \mathbf{U}^2_3 \xleftrightarrow{R_2} \mathbf{0}; \quad (g) \tag{2.17}$$

Remark 2.2. Note that although $\mathbf{V}^j_{i(\bar{n})}$ and $\bar{\mathbf{V}}^j_{i(\bar{n})}$ are designed in the similar way, they are in fact independent of each other almost for sure. For example, let $\mathbf{V}^1_{i(1)}$ and $\mathbf{V}^1_{i(2)}$ are designed from the original chain with $S = 4$, i.e.,

$$\mathbf{0} \xleftrightarrow{R_2} \mathbf{V}^1_{1(1)} \xleftrightarrow{R_3} \mathbf{V}^1_{2(1)} \xleftrightarrow{R_1} \mathbf{V}^1_{3(1)} \xleftrightarrow{R_2} \mathbf{V}^1_{1(2)} \xleftrightarrow{R_3} \mathbf{0} \tag{2.18}$$

and $\bar{\mathbf{V}}_{i(1)}^1$, $\bar{\mathbf{V}}_{i(2)}^1$ are designed from the long chain as follows,

$$\mathbf{0} \xleftrightarrow{R_2} \bar{\mathbf{V}}_{1(1)}^1 \xleftrightarrow{R_3} \bar{\mathbf{V}}_{2(1)}^1 \xleftrightarrow{R_1} \bar{\mathbf{V}}_{3(1)}^1 \xleftrightarrow{R_2} \bar{\mathbf{V}}_{1(2)}^1 \xleftrightarrow{R_3} \bar{\mathbf{V}}_{2(2)}^1 \xleftrightarrow{R_1} \mathbf{0} \qquad (2.19)$$

(2.18) shows that $\mathbf{V}_{1(2)}^1$ is nulled on receiver 3, i.e., $\mathbf{H}_{13}\mathbf{V}_{1(2)}^1 = \mathbf{0}$, whereas $\bar{\mathbf{V}}_{1(2)}^1$ in (2.19) is not nulled on receiver 3, i.e., $\mathbf{H}_{13}\bar{\mathbf{V}}_{1(2)}^1 \neq \mathbf{0}$, which means $\mathbf{V}_{1(2)}^1$ and $\bar{\mathbf{V}}_{1(2)}^1$ are independent of each other for sure. Accordingly, it can be easily proved that $\mathbf{V}_{3(1)}^1$ and $\bar{\mathbf{V}}_{3(1)}^1$ are independent of each other. Hence, it can be seen that based on this design, $\mathbf{V}_{i(\bar{n})}^j$ and $\bar{\mathbf{V}}_{i(\bar{n})}^j$ are independent of each other.

2.2.4 Signal Processing at Users

After presenting the design of beamforming at BSs, we now explain the signal processing strategy at the users.

Based on (2.6), the received signals on user i can be written as

$$\mathbf{y}_i = \mathbf{H}_{ii}\mathbf{B}_i\mathbf{m}_i + \underbrace{\begin{bmatrix} \mathbf{H}_{ij}\mathbf{B}_j & \mathbf{H}_{ik}\mathbf{B}_k \end{bmatrix}}_{\mathbf{I}_i} \begin{bmatrix} \mathbf{m}_j \\ \mathbf{m}_k \end{bmatrix} + \mathbf{z}_i \qquad (2.20)$$

where $i \neq j \neq k$ and $\mathbf{I}_i \in \mathbb{C}^{M_R \times (d_j + d_k)}$ denotes the space of inter-cell interference at user i.

Then, a zero-forcing filter $\mathbf{U}_i \in \mathbb{C}^{M_R \times d_i}$ should be designed to null the interference $\mathbf{I}_i$, i.e.,

$$\mathbf{U}_i^H\mathbf{I}_i = \mathbf{0} \qquad (2.21)$$

Note that since user i has d_i desired messages, i.e., $\mathbf{m}_i \in \mathbb{C}^{d_i \times 1}$, $\mathbf{U}_i$ must have full rank of d_i so that all desired messages can be linearly separated. Further, based on the design in Sect. 2.2.3, the rank of $\mathbf{I}_i$ must be less than $M_R - d_i$. Hence, a qualified $\mathbf{U}_i$ can be found in (2.21) for sure.

Consequently, by applying the filter $\mathbf{U}_i'$ at the received signal, we have

$$\mathbf{y}_i' = \mathbf{U}_i'\mathbf{y}_i = \mathbf{m}_i + \mathbf{U}_i'\mathbf{z}_i \qquad (2.22)$$

where

$$\mathbf{U}_i' = (\mathbf{U}_i^H\mathbf{H}_{ii}\mathbf{B}_i)^{-1}\mathbf{U}_i^H \qquad (2.23)$$

Note that since the BSs have global CSI, the filter $\mathbf{U}_i'$ can be designed at the BS i according to (2.21), and then forwarded to user i through an interference-free, high signal-to-noise ratio (SNR) control link [35–38].

2.3 Cloud Radio Access Network

As we can see, with CSI-sharing BS cooperation, some resources are usually sacrificed to mitigate the interference. Hence, more powerful cooperation is needed so that the channel resources can be fully exploited and the performance of the network can be further improved in terms of both spectral efficiency and energy efficiency.

In this section, we introduce a recently proposed network architecture, called C-RAN, which is regarded as a potential ultimate solution to the "spectrum crunch" of cellular networks [39–41]. In the following, we provide a brief review of C-RAN from the perspective of system structure and key design challenges.

2.3.1 System Structure

C-RAN has been proposed as a combination of emerging technologies from both the wireless and the information technology industries by incorporating cloud computing into radio access networks (RANs) [39]. In C-RAN, the traditional role of the BSs is decoupled into two parts: the distributed installed remote radio heads (RRHs) and the baseband units (BBUs) clustered as a BBU pool in a centralized cloud server, as shown in Fig. 2.5, which is on top of next page.

RRHs support seamless coverage and provide high capacity in hot spots, while BBUs provide large-scale processing and management of the signals from/to RRHs, where cloud computing technology provides flexible spectrum management and advanced network coordination [41]. By taking advantages of the high-capacity backhual links between RRHs and BBU pool, the C-RAN architecture enables joint encoding and decoding of messages from multiple cells, which means the virtual MIMO BS cooperation can be realized.

Remote Radio Head

Since most of the computation resources are aggregated in the BBU pool, the functions of RRH can be relatively simple. The basic function of RRH is to transmit radio frequency signals to user equipments (UEs) in the downlink and forward the baseband signals from UEs to the BBU pool in the uplink. The advantage of such simple low-cost RRH is that they can be easily deployed and installed in a very high density in large scale scenarios.

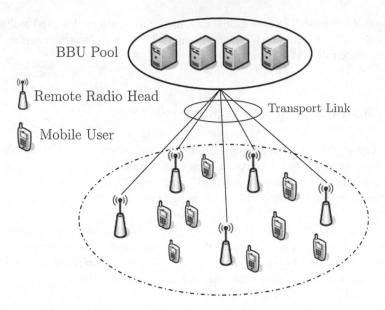

Fig. 2.5 The architecture of cloud RAN

BBU Pool

A BBU pool is located at a centralized site and consists of time-varying sets of software defined BBUs, which operate as virtual BSs to process baseband signals and optimize the radio resource allocation [41]. At BBU pool, the CSI, traffic data, and control information of mobile services can be fully shared. As a result, a virtual MIMO system can be formed from BBU pool's perspective, which enables multiplexing more streams on the same channel without mutual interference.

Fronthaul

Fronthaul is defined as the links between RRHs and BBU pools. The two parts can be connected via either optical fibers or standard wireless channels. Optical fiber transmission is considered to be ideal fronthauls as it provides high transmission capacity and very short delays. However, it will largely increase the cost and lead to inflexible deployment. On the other hand, if the transmission is through wireless channels, the limits of capacity and other constraints must be taken into consideration.

2.3.2 Design Challenges

The new architecture of C-RAN indicates a paradigm shift in the network design, which will certainly cause some technical challenges for implementations. In the following, we discuss some key design issues of C-RAN.

Information Compression

A prerequisite requirement for complete centralized processing in the BBU pool is an ideal fronthaul with high bandwidth and low time latency. However, as mentioned before, due to high cost of optical fiber, the fronthaul in practice is often capacity constrained or time-delay constrained [42]. Therefore, to overcome the impact of constrained fronthaul, signal compression is required.

In the uplink of a C-RAN, each RRH compresses the received signals and relay the "soft" information to the cloud decoder [42, 43]. Since the signals received at different RRHs are correlated, distributed source coding strategies can be used to design robust and efficient compression scheme [42]. In addition, in C-RAN, joint decompression and decoding can be performed at cloud decoder [44].

As a counterpart of the distributed source coding strategy for the uplink, the joint design of precoding and compression for the finite capacity fronthaul in downlink is studied in [45].

Precoding

Large-scale coordinated precoding in BBU pool is a promising technique to improve network performance. Both spectrum efficiency and energy efficiency should be taken into consideration. Note that in C-RAN, each user is only close to a few RRHs in its neighborhood, and vice versa. Therefore, the channel matrix should only contains a small fraction of entries with large enough gains. Thus, ignoring the small entries in the channel matrix can introduce sparsity to the matrix, which is a desirable property for C-RAN as it can significantly reduce the processing complexity and energy consumption of the network [40].

Resource Allocation

In fronthaul constrained C-RAN, advanced resource allocation and optimization is often required due to the densely distributed RRH and powerful centralized BBUs. Multi-dimensional joint resource optimization including precoding, user scheduling, resource allocation, and cell association can significantly enhance the overall system performance [41].

In addition, clustering is another important technique in C-RAN that is often used to reduce the channel estimation overhead and computational complexity. In large scale C-RAN, a typical user is only served by the RRHs in the same cluster. There are generally two types of clustering schemes for data sharing, which are disjoint clustering and user-centric clustering. In disjoint clustering scheme, the entire network is divided into non-overlapping clusters and the RRHs in each cluster can jointly serve all the users within the area. With disjoint clustering scheme, the users at the edge of clusters may suffer severe inter-cluster interference. Alternatively, in user centric clustering, each user is served by an individually selected subset of RRHs and different clusters for different users may overlap. With user-centric clustering, there exists no explicit cluster edge [46].

2.4 Summary

In this Chapter, we discussed the interference coordination techniques for multi-cell MIMO networks. Specifically, we focus on the users in the cell-edge area that may suffer interference from other cells. Both CSI-sharing BS cooperation and MIMO BS cooperation are considered. With CSI-sharing cooperation, a practical downlink scenario was investigated, where three users in the intersection of three cells share the same time-frequency resources, and each user belongs to one cell. A spatial beamforming scheme was proposed to coordinate the inter-cell interference. The performance of the scheme was evaluated in terms of degrees of freedom (DoF). It was shown that the proposed scheme can achieve the optimal *DoF region*. Then, we introduced cloud radio access network (C-RAN), which can be seen as a platform for the implementation of virtual MIMO BS cooperation. The system structure and key design challenges of C-RAN were presented.

References

1. Wu, J., Zhang, Z., Hong, Y., & Wen, Y. (2015). Cloud radio access network (C-RAN): A primer. *IEEE Network, 29*(1), 35–41.
2. Jafar, S., & Fakhereddin, M. (2007). Degrees of freedom for the MIMO interference channel. *IEEE Transactions on Information Theory, 53*(7), 2637–2641.
3. Gou, T., & Jafar, S. (2010). Degree of freedom of the K-user $M \times N$ MIMO interference channel. *IEEE Transactions on Information Theory, 56*(12), 6040–6057.
4. Shin, W., Lee, N., Lim, J., Shin, C., & Jang, K. (2011). On the design of interference alignment scheme for two-cell MIMO interference broadcast channels. *IEEE Transactions on Wireless Communications, 10*, 437–442.
5. Suh, C., Ho, M., Lim, J., & Tse, D. (2011). Downlink interference alignment. *IEEE Transactions on Communications, 59*(9), 2616–2626.
6. Yang, L., & Zhang, W. (2013). Opportunistic interference alignment in heterogeneous two-cell uplink network. In *Proceedings of the IEEE International Conference on Communications (ICC 2013)*, Budapest, pp. 5448–5452, 9–13 June 2013.

7. Cadambe, V., & Jafar, S. (2008). Interference alignment and degrees of freedom of the K-user interference channel. *IEEE Transactions on Information Theory, 54*(8), 3425–3441.

8. Ghasemi, A., Motahari, A., & Khandani, A. (2010). Interference alignment for the K-user MIMO interference channel. In *Proceedings of the IEEE International Symposium on Information Theory (ISIT)*, Austin, pp. 360–364, 13–18 June 2010.

9. Yetis, C., Gou, T., Jafar, S., & Kayran, A. (2010). On feasibility of interference alignment in MIMO interference network. *IEEE Transactions on Signal Processing, 58*(9), 4771–4782.

10. Bresler, G., Cartwright, D., & Tse, D. (2014). Feasibility of interference alignment for the MIMO interference channel. *IEEE Transactions on Information Theory, 60*(9), 5573–5586.

11. Razaviyayn, M., Lyubeznik, G., & Luo, Z. (2012). On the Degrees of freedom achievable through interference alignment in a MIMO interference channel. *IEEE Transactions on Signal Processing, 60*(2), 812–821.

12. Razaviyayn, M., Sanjabi, M., & Luo, Z. (2012). Linear transceiver design for interference alignment: Complexity and computation. *IEEE Transactions on Information Theory, 58*(5), 2896–2910.

13. Wang, C., Gou, T., & Jafar, S. (2014). Subspace alignment chains and the degree of freedom of the three-user MIMO interference channel. *IEEE Transactions on Information Theory, 60*(5), 2432–2479.

14. Bresler, G., Cartwright, D., & Tse, D. (2011). Geometry of the 3-user MIMO interference channel. In *Proceedings of the Allerton Conference on Communication, Control, and Computing*, Monticello, pp. 1264–1271, 28–30 Sept 2011.

15. Tse, D., & Viswanath, P. (2005). *Fundamentals of wireless communications*. Cambridge: Cambridge University Press.

16. Jafar, S., & Shamai, S. (2008). Degrees of freedom region for the MIMO X channel. *IEEE Transactions on Information Theory, 54*(1), 151–170.

17. Gomadam, K., Cadambe, V., & Jafar, S. (2011). A distributed numerical approach to interference alignment and applications to wireless interference networks. *IEEE Transactions on Information Theory, 57*(6), 3309–3322.

18. Yang, L., & Zhang, W. (2014). Interference alignment with asymmetric complex signaling on MIMO X channels. *IEEE Transactions on Communications, 62*(10), 3560–3570.

19. Maddah-Ali, M., Motahari, A., & Khandani, A. (2006). Signaling over MIMO multibase systems: Combination of multiaccess and broadcast schemes. In *Proceedings of the IEEE International Symposium on Information Theory (ISIT)*, Seattle, pp. 2104–2108, 9–14 July 2006.

20. Bresler, G., Parekh, A., & Tse, D. (2010). The approximate capacity of the many-to-one and one-to-many Gaussian interference channel. *IEEE Transactions on Information Theory, 56*(9), 4566–4580.

21. Cadambe, V., Jafar, S., & Shamai, S. (2009). Interference alignment on the deterministic channel and application to Gaussian networks. *IEEE Transactions on Information Theory, 55*(1), 269–274.

22. Motahari, A., Oveis-Gharan, S., Maddah-Ali, M., & Khandani, A. (2014). Real interference alignment: Exploiting the potential of single antenna systems. *IEEE Transactions on Information Theory, 60*(8), 4799–4810.

23. Etkin, R., & Ordentilich, E. (2009). The degrees of freedom of the K-user Gaussian interference channel is discontinuous at rational channel coefficients. *IEEE Transactions on Information Theory, 55*(11), 4932–4946.

24. Avestimehr, S., Diggavi, S., & Tse, D. (2007). Determinister approach to wireless relay networks. In *Proceedings of the Allerton Conference on Communication, Control, and Computing*, Monticello, Sept 2007

25. Perlaza, S. M., Fawaz, N., Lasaulce, S., & Debbah, M. (2010). From spectrum pooling to space pooling: Opportunistic interference alignment in MIMO cognitive networks. *IEEE Transactions on Signal Processing, 58*(7), 3728–3741.

26. Amir, M., El-Keyi, A., & Nafie, M. (2010). Opportunistic interference alignment for multiuser cognitive radio. In *Proceedings of the IEEE Information Theory Workshop*, Cairo, 6–8 Jan 2010.

27. Yang, L., Zhang, W., Zheng, N., & Ching, P. C. (2014). Opportunistic user scheduling in MIMO cognitive radio networks. In *Proceedings of the IEEE International Conference on Acoustics, Speech and Signal Processing (ICASSP 2014)*, Florence, pp 7303–7307, 4–9 May 2014.
28. Gou, T., Jafar, S., Wang, C., Jeon, S., & Chung, S. (2012). Aligned interference neutralization and the degrees of freedom of the $2 \times 2 \times 2$ interference channel. *IEEE Transactions on Information Theory, 58*(7), 4381–4395.
29. Wang, Z., Xiao, M., Wang, C., & Skoglund, M. (2013). Degrees of freedom of multi-hop MIMO broadcast networks with delayed CSIT. *IEEE Communications Letters, 2*(2), 207–210.
30. Yang, L., & Zhang, W. (2014). Degrees of freedom of relay-assisted MIMO interfering broadcast channels. In *Proceedings of the IEEE Globe Communications Conference (Globecom)*, Austin, 8–12 Dec 2014.
31. Lee, N., & Heath, R., Jr. (2013). Degrees of freedom for the two-cell two-hop MIMO interference channel: Interference-free relay transmission and spectrally efficient relaying protocol. *IEEE Transactions on Information Theory, 59*(5), 2882–2896.
32. Yang, L., Zhang, W., & Jin, S. (2015). Interferene alignment in device-to-device LAN underlaying cellular network. *IEEE Transactions on Wireless Communications, 14*(7), 3715–3723.
33. Cadambe, V., Jafar, S., Maleki, H., Ramchandran, K., & Suh, C. (2013). Asymptotic interference alignment for optimal repair of MDS codes in distributedstorage. *IEEE Transactions on Information Theory, 59*(5), 2974–2987.
34. Yang, L., & Zhang, W. (2015). On degrees of freedom region of three-user MIMO interference channels. *IEEE Transactions on Signal Processing, 63*(3), 590–603.
35. Jung, B., & Shin, W. (2011). Opportunistic interference alignment for interference-limited cellular TDD uplink. *IEEE Communications Letters, 15*(2), 148–150.
36. Cho, S., Huang, K., Kim, D., Lau, V., Chae, H., Seo, H., & Kim, B. (2012). Feedback-topology designs for interference alignment in MIMO interference channels. *IEEE Transactions on Signal Processing, 60*(12), 6561–6575.
37. Rao, X., Ruan, L., & Lau, V. (2013). Limited feedback design for interference alignment on MIMO interference networks with heterogeneous path loss and spatial correlations. *IEEE Transactions on Signal Processing, 61*(10), 2598–2607.
38. Chae, C., Inoue, T., Mazzarese, D., & Heath, R., Jr. (2008). Coordinated beaforming for the multiuser MIMO broadcast channel with limited feedforward. *IEEE Transactions on Signal Processing, 56*(12), 6044–6056.
39. Chih-Lin, I., Rowell, C., Han, S., Xu, Z., Li, G., & Pan, Z. (2014). Toward green and soft: A 5G perspective. *IEEE Communications Magazine, 52*(2), 66–73.
40. Shi, Y., Zhang, J., & Letaief, K. B. (2014). Group sparse beamforming for green cloud-RAN. *IEEE Transactions on Wireless Communications, 13*(5), 2809–2823.
41. Peng, M., Wang, C., Lau, V., & Poor, H. V. (2015). Fronthaul-constrained cloud radio access networks: Insights and challenges. *IEEE Transactions on Wireless Communications, 22*(2), 152–160.
42. Park, S., Simeone, O., Sahin, O., & Shamai, S. (2013). Robust and efficient distributed compression for cloud radio access networks. *IEEE Transactions on Vehicular Technology, 62*(2), 692–703.
43. Rao, X., & Lau, V. (2015). Distributed fronthaul compression and joint signal recovery in cloud-RAN. *IEEE Transactions on Signal Processing, 63*(4), 1056–1065.
44. Park, S., Simeone, O., Sahin, O., & Shamai, S. (2013). Joint decompression and decoding for cloud radio access networks. *IEEE Transactions on Signal Processing Letter, 20*(5), 503–506.
45. Park, S., Simeone, O., Sahin, O., & Shamai, S. (2013). Joint precoding and multivariate backhaul compression for the downlink of cloud radio access networks. *IEEE Transactions on Signal Processing, 20*(5), 503–506.
46. Dai, B., & Yu, W. (2014). Sparse beamforming and user-centric clustering for downlink cloud radio access network. *IEEE Access, 2*, 1326–1339.

Chapter 3
Interference Coordination in Device-to-Device Communication

3.1 Introduction

Device-to-device (D2D) communications enable nearby mobile devices to establish direct links in cellular networks without traversing the core network [1, 2]. Exploiting direct communications between user equipments (UEs) will improve spectrum utilization, shorten packet delay, reduce energy consumption, and enable new peer-to-peer location-based applications and services. Moreover, D2D-enabled devices can also be a required feature in public safety network [3].

The D2D communications can occur in either licensed cellular spectrum (in-band) or unlicensed spectrum (out-band). There are two key problems for in-band D2D communications: 1. How can a UE become a potential D2D user (DU); 2. How does a DU access the licensed spectrum of cellular networks. The first problem is regarding to the initial phase of the setup of D2D links, where the UEs in close proximity are paired. This is accomplished during neighbor discovery process, where UEs will identify their neighbors for potential direct-link setup [4]. Besides the distance between two users, other factors including instantaneous network load, channel conditions, and interference situations should all be taken into account when selecting potential DUs [5].

In this chapter, we investigate the second problem, i.e., how DUs access the spectrum of cellular network. In general, there are two ways for DUs to access the spectrum, which are overlay communications and underlay communications. In overlay D2D transmission, the spectrum is divided into two orthogonal portions. A fractional η is assigned to D2D communications while the other $1 - \eta$ is used for

© [2015] IEEE. Reprinted, with permission, from [L. Yang, W. Zhang and S. Jin, "Interference alignment in device-to-device LAN underlaying cellular networks," *IEEE Trans. Wireless Comm.*, vol. 14, pp. 3715–3723, July 2015.]

© The Author(s) 2015
L. Yang, W. Zhang, *Interference Coordination for 5G Cellular Networks*,
SpringerBriefs in Electrical and Computer Engineering,
DOI 10.1007/978-3-319-24723-6_3

cellular communication. The value of η should be optimized in terms of spectrum efficiency or energy efficiency [6]. Overlay approaches eliminate the concerns for interference from D2D communications on cellular transmissions, but reduce the amount of available resources for cellular communications [7].

The coexistence of D2D and cellular communications in the same spectrum poses many challenges and risks due to the difficulty of interference management. In particular, the underlaid D2D signals become a new source of interference. Consequently, cellular links experience cross-tier interference from D2D transmissions while the D2D links suffer from not only the cross-tier interference from cellular transmissions but also the inter-D2D interference. Hence, an effective interference coordination mechanism is needed to ensure successful coexistence of cellular and D2D links. Currently, the majority of works deal with the problem from the aspects of resource scheduling and power control. Many schemes are proposed to jointly allocate the physical resource blocks (in both time domain and frequency domain) and perform power control for D2D links subject to interference constraints for cellular operations and quality of services (QoS) demands of D2D links [8–15].

In modern cellular networks, devices are usually equipped with multiple antennas [16, 17]. MIMO techniques have been widely used to improve the performance of communications. Further, beamforming based on multi-antennas has been identified as a key technique to coordinate the interference in wireless networks [18–20]. It is expected that D2D transceivers can use multiple antennas and MIMO schemes to increase data throughput and link coverage without increased transmission power or bandwidth [21]. Since beamforming is performed in spatial domain, it deals with the interference among users in the same time-frequency resource blocks. Hence, it can be used as a compliment of existing resource allocation schemes and further improve the spectrum efficiency. Some works have already been done to manage the interference of D2D underlaid cellular networks with beamforming techniques [22–28]. In [22], a downlink MIMO beamforming scheme was proposed for base station (BS) to transmit signals in the null space of the interference between BS and D2D receiver (DR). In [23], two beamforming schemes for BS were proposed. The first one is similar to [22], where BS uses beamforming to cancel the interference to DR. In the second one, the BS aims at serving its cellular user (CU) rather than dealing with interference. The capacity performance of these two cases were evaluated in [23]. The same system model was also studied in [24], where jointly precoding on both BS and D2D transmitter is used to maximize the signal-to-leakage-noise ratio (SLNR) or signal-to-interference-noise ratio (SINR). In [25], linear precoder-decoder schemes are proposed for two-way relaying based D2D communications, where the relay uses physical layer network coding. In the work of [22–25], one cellular link shares the same resource blocks with only one D2D link. In [26], a beamforming scheme based on interference alignment (IA) was proposed for three D2D links, but their effects on cellular network were not studied. In [27], IA was adopted to improve the energy efficiency for both D2D and cellular links. Further, the interplay between massive MIMO and D2D networking was studied in [28].

In this Chapter, we consider a D2D local area networks (LAN) underlaying a cellular uplink, where multiple DUs intend to communicate with a D2D receiver. This model can be found in many practical scenarios. For example, some context-aware applications (on cell phone) allow nearby devices to discover each other and exchange messages directly. In some occasions, there are many devices with same application gathering together, and they need to communicate with a common DU. Two interference coordination schemes for D2D communications are proposed to manage the interference between the two networks for different scenarios. The first scheme is referred to as 'interference-free' scheme, which can be applied in the scenarios where some sub-channels of BS are not occupied by CUs. In this scheme, the interference signals from DUs are aligned in the orthogonal space of cellular links at the BS. Hence, the links of CUs are completely free from interference. Note that if all the channel spaces of BS are used by CUs, the orthogonal space of cellular links may not exist. In case of such scenarios, we propose another scheme which is referred to as 'interference-limiting' scheme. In this scheme, the DUs' signals are allowed to occupy some links of CUs, but the peak interference power on each of the 'interfered' links is kept under a certain threshold γ. It is shown that the 'interference-limiting' scheme is most efficient for the scenarios where there are a large number of DUs. The explicit design frameworks and feasibility conditions of the two schemes are provided. Performance analysis shows that based on the proposed schemes, the interference generated on the cellular links is eliminated or well controlled, while the QoS of the D2D LAN can also be guaranteed.

The rest of the Chapter is organized as follows. In Sect. 3.2, system model is introduced. In Sect. 3.3, the 'interference-free' D2D communication scheme is proposed, followed by the performance analysis of the networks. In Sect. 3.4, the 'interference-limiting' D2D communication scheme is proposed, followed by the performance analysis of the networks. In Sect. 3.5, simulation results are presented and discussed. Section 3.6 summarizes the Chapter.

3.2 System Model

We consider a D2D LAN underlaying a cellular uplink in a single cell setting, where multiple DUs intend to communicate with a DR, as shown in Fig. 3.1. The BS receives both desired signals and interference signals from CUs and DUs, respectively. The BS and DR are equipped with N and N_d antennas, respectively. The CU and DU each are equipped with M and M_d antennas, respectively.

Let $\mathbf{H}_{CU,i}^e$ denote the channel from CU_i to BS, and $\mathbf{u}_i$ denote the precoding vector of CU_i. The received signals on BS at the absence of interference can be expressed as

$$\mathbf{y}_e' = \mathbf{D}_e \mathbf{y}_e = \mathbf{D}_e \sum_{i=1}^{s} \mathbf{H}_{CU,i}^e \mathbf{u}_i P_c^{\frac{1}{2}} m_i + \mathbf{D}_e \mathbf{z}_e$$

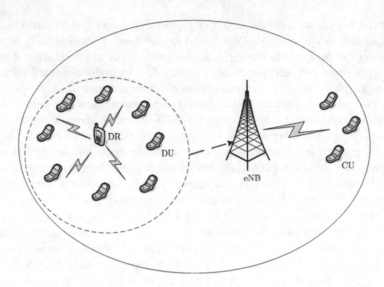

Fig. 3.1 D2D LAN underlaying a cellular uplink network

$$= P_c^{\frac{1}{2}} \mathbf{D}_e \begin{bmatrix} \mathbf{H}_{CU,1}^e \mathbf{u}_1 & \cdots & \mathbf{H}_{CU,s}^e \mathbf{u}_s \end{bmatrix} \begin{bmatrix} m_1 \\ \vdots \\ m_s \end{bmatrix} + \mathbf{D}_e \mathbf{z}_e$$

where $\mathbf{D}_e \in \mathbb{C}^{s \times N}$ denotes the post-processing matrix of BS, s denotes the number of links of CU and $s \leq N$, m_i denotes the message of CU_i, P_c denotes the transmit power of CU, $\mathbf{z}_e$ denotes the noise on BS with unit variance. The design of $\mathbf{D}_e$ is different in two schemes, which will be explained later.

Then, $\mathbf{u}_i$ can be designed to lie in the null space of $\overline{\mathbf{D}_e \mathbf{H}_{CU,i}^e(i)}$, i.e.,

$$\overline{\mathbf{D}_e \mathbf{H}_{CU,i}^e(i)} \cdot \mathbf{u}_i = \mathbf{0} \tag{3.1}$$

where $\overline{\mathbf{D}_e \mathbf{H}_{CU,i}^e(i)}$ denotes the matrix $\mathbf{D}_e \mathbf{H}_{CU,i}^e$ without the ith row.

As a result, the links from CUs to BS can be transformed to s parallel links, i.e.,

$$\mathbf{y}_e' = P_c^{\frac{1}{2}} \underbrace{\begin{bmatrix} \lambda_1 & & \\ & \ddots & \\ & & \lambda_s \end{bmatrix}}_{\psi \in \mathbb{C}^{s \times s}} \begin{bmatrix} m_1 \\ \vdots \\ m_s \end{bmatrix} + \mathbf{z}_e' \tag{3.2}$$

where $\lambda_i = \mathbf{D}_e \mathbf{H}_{CU,i}^e(i) \mathbf{u}_i$ denotes the equivalent channel gain of the i-th link of CU, and $\mathbf{D}_e \mathbf{H}_{CU,i}^e(i)$ denotes the i-th row of $\mathbf{D}_e \mathbf{H}_{CU,i}^e$. Since there are totally N antennas on BS, there will be $r = N - s$ dimensions left 'unused' by CUs.

Let $\mathbf{H}^e_{DU,i}$ denote the channel from DU_i to BS and $\mathbf{v}_i$ denote the beamforming vector of DU_i. Assuming there are l active DUs, the received signal on BS in the presence of interference becomes

$$\mathbf{y}_e = P_c^{\frac{1}{2}}\boldsymbol{\psi}\mathbf{m} + \mathbf{D}_e \sum_{i=1}^{l} \mathbf{H}^e_{DU,i}\mathbf{v}_i P_d^{\frac{1}{2}} t_i + \mathbf{z}'_e \tag{3.3}$$

where $\mathbf{m} = \begin{bmatrix} m_1 & \cdots & m_s \end{bmatrix}^T$, t_i denotes the message of DU_i and P_d denotes the power of t_i, and the design of $\mathbf{v}_i$ is specified in different schemes.

Then, the received signal on DR is given by

$$\mathbf{y}_r = \sum_{i=1}^{l} \mathbf{H}^r_{DU,i}\mathbf{v}_i P_d^{\frac{1}{2}} t_i + \sum_{i=1}^{s} \mathbf{H}^r_{CU,i}\mathbf{u}_i P_c^{\frac{1}{2}} m_i + \mathbf{z}_r \tag{3.4}$$

where $\mathbf{H}^r_{DU,i}$ and $\mathbf{H}^r_{CU,i}$ denote the channels from DU_i and CU_i to DR, respectively.

$$\underbrace{\begin{bmatrix} a \cdot \mathbf{H}^e_{DU,1} & b \cdot \mathbf{H}^e_{DU,2} & f \cdot \mathbf{H}^e_{DU,r} & -\mathbf{H}^e_{DU,r+1} & 0 & 0 & 0 \\ g \cdot \mathbf{H}^e_{DU,1} & \varepsilon \cdot \mathbf{H}^e_{DU,2} & q \cdot \mathbf{H}^e_{DU,r} & 0 & -\mathbf{H}^e_{DU,r+2} & 0 & 0 \\ \vdots & \vdots & \cdots & \vdots & \vdots & \vdots & \ddots & \vdots \\ \gamma \cdot \mathbf{H}^e_{DU,1} & \kappa \cdot \mathbf{H}^e_{DU,2} & \lambda \cdot \mathbf{H}^e_{DU,r} & 0 & 0 & 0 & -\mathbf{H}^e_{DU,l} \end{bmatrix}}_{\mathbf{H}' \in \mathbb{C}^{N \cdot (l-r) \times M_d \cdot l}} \begin{bmatrix} \mathbf{v}'_1 \\ \vdots \\ \mathbf{v}'_r \\ \mathbf{v}'_{r+1} \\ \vdots \\ \mathbf{v}'_l \end{bmatrix} = \mathbf{0}$$

$$\tag{3.5}$$

where $a \cdots \lambda$ are all arbitrary complex parameters.

Further, DU_i only knows the channel $\mathbf{H}^e_{DU,i}$. All the channel matrices are sampled from continuous complex Gaussian distributions and each entry is independent and identically distributed (i.i.d.) with zero mean and unit variance. The channels are assumed to undergo block fading.

Finally, from (3.3) we should note that if $s < M_d$, then $\mathbf{D}_e\mathbf{H}^e_{DU,i} \in \mathbb{C}^{s \times M_d}$ is a 'fat' matrix, which means $\mathbf{v}_i$ can be found to null the interference at BS almost surely, regardless of the choice of $\mathbf{D}_e$. Hence, in the following, we only focus on the case of $s \geq M_d$.

3.3   Interference-Free D2D Communication Scheme

In this section, we introduce the 'interference free' IA scheme for D2D communications. We first elaborate the design process of the scheme. Then, the performance analysis of both D2D LAN and cellular link is provided. Finally, the advantage and limitation of this scheme are discussed.

3.3.1 Design Process

Four steps are involved in the design process of D2D transmission. In **Step 1**, the BS selects l DUs for D2D communications, where $l < \frac{N \cdot r}{N - M_d}$ and $l \le N_d$. In **Step 2**, the post-processing matrix $\mathbf{D}_e$ is designed and broadcasted through physical downlink control channels. In **Step 3**, each DU designs the precoding vectors $\mathbf{v}_i$ according to the received $\mathbf{D}_e$. Each CU also designs precoding vector $\mathbf{u}_i$ as described in Sect. 3.2. In **Step 4**, the DR designs the receiving filter to decode transmitted signals.

- **Step 1:** The l DUs can be selected randomly, or according to some protocols or preference of DR. To ensure that the signals from DUs can be linearly decoded at DR, it must have $l \le N_d$. Further, we have following result.

Theorem 3.1. *In a cellular uplink network underlaying a D2D LAN, where there are s links of CUs and l active DUs, the interference on BS can be nulled while the desired signals can still be linearly decoded as long as $l < \frac{N \cdot r}{N - M_d}$, where $r = N - s$.*

Proof. The BS receives s desired signals and l interference signals. Since the desired signals occupy s dimensions, the l interference signals must be aligned to $r = N - s$ dimensions. To do so, we should align each of the last $l - r$ interference signals within the interference space that is spanned by the first r interference signals, as shown in (3.5).

As we can see, the non-zero solution for $\left[\mathbf{v}'_1 \ \cdots \ \mathbf{v}'_l \right]^T$ can be found if $\mathbf{H}'$ is a 'fat' matrix, i.e.,

$$M_d \cdot l > N \cdot (l - r) \tag{3.6}$$

which is equivalent to

$$l < \frac{N \cdot r}{N - M_d} \tag{3.7}$$

Moreover, since all the channels are generic, there are at least $M_d \cdot l + 1$ non-zero entries in $\left[\mathbf{v}'_1 \ \cdots \ \mathbf{v}'_l \right]^T$. Hence, each of the l precoding vectors can be guaranteed to be non-zero vector almost surely.

- **Step 2:** After selecting the potential DUs, the BS needs to design a proper post processing matrix $\mathbf{D}_e$, which should be orthogonal to the interference space. Since the BS knows all the receiving channels, it can formulate equation (3.5) with the channel matrices between selected DUs and BS. Hence, the interference space that formed by DUs' signals is $\text{span}\{\left[\mathbf{H}^e_{DU,1} \mathbf{v}'_1 \ \cdots \ \mathbf{H}^e_{DU,r} \mathbf{v}'_r \right]\} \in \mathbb{C}^{N \times r}$. Then, $\mathbf{D}_e \in \mathbb{C}^{(N-r) \times N}$ can be designed as follows,

$$\mathbf{D}_e \cdot \left[\mathbf{H}^e_{DU,1} \mathbf{v}'_1 \ \cdots \ \mathbf{H}^e_{DU,r} \mathbf{v}'_r \right] = \mathbf{0} \tag{3.8}$$

Moreover, we assume that BS will send $\mathbf{D}_e$ to all users (DUs and CUs) through the physical downlink control channels [29, 30]. Specifically, the matrix $\mathbf{D}_e$ can be sent in an unquantized uncoded version based on the analog channel state information (CSI) feedback scheme proposed in [31, 32]. Alternatively, $\mathbf{D}_e$ can also be transmitted with conventional CSI feedback mechanism based on quantization codebooks [33–35]. First, a common codebook can be stored in BS and users. Then, $\mathbf{D}_e$ can be quantized by using the codebook. Each time, the eNB will only broadcast the index of the corresponding codeword of $\mathbf{D}_e$.

- **Step 3:** After receiving $\mathbf{D}_e$, each DU can design the precoding vector accordingly.

  The DU_i first calculates $\mathbf{P}_i \in \mathbb{C}^{(N-r) \times M_d} = \mathbf{D}_e \mathbf{H}_{DU,i}^e$ (Note that $\mathbf{H}_{DU,i}^e$ is known by DU_i). Then, the precoding vector $\mathbf{v}_i$ is determined by calculating the null space of $\mathbf{P}_i$, i.e,

$$\mathbf{P}_i \mathbf{v}_i = \mathbf{0} \tag{3.9}$$

Since $N - r = s \leq M_d$, $\mathbf{P}_i$ is either a square or 'thin' matrix, which means its null space does not exist if $\mathbf{P}_i$ is full rank.

If the DU_i is one of DUs that are selected by the BS in **Step 1**, $\mathbf{H}_{DU,i}^e$ is one of the channel matrices in (3.5), which means

$$\mathbf{H}_{DU,i}^e \mathbf{v}_i' = \left[\mathbf{H}_{DU,1}^e \mathbf{v}_1' \cdots \mathbf{H}_{DU,r}^e \mathbf{v}_r' \right] \begin{bmatrix} a_1 \\ \vdots \\ a_r \end{bmatrix}. \tag{3.10}$$

If $r + 1 \leq i \leq l$, $a_1 \cdots a_r$ are all random complex numbers; if $1 \leq i \leq r$, $a_1 \cdots a_r$ are all zeros except $a_i = 1$.

Hence, based on (3.8), we can see that $\mathbf{P}_i$ must not be full rank and its null space is $\mathbf{v}_i = \text{span}\{\mathbf{v}_i'\}$. Therefore, by calculating (3.9), DU_i can find the precoding vector as $\mathbf{v}_i = \text{span}\{\mathbf{v}_i'\}$.

On the other hand, if the DU_i is not selected by the BS, then $\mathbf{P}_i$ is just a random matrix that has full rank almost surely. Therefore, since a non-zero solution of $\mathbf{v}_i$ that satisfies (3.9) does not exist, those unselected DUs will not start the D2D transmission.

- **Step 4:** Finally, DR designs its post processing matrix $\mathbf{U}_r$ to decode signals from DUs. Let $\mathfrak{g}_i \in \mathbb{C}^{N_d \times 1} = \mathbf{H}_{DU,i}^r \mathbf{v}_i$ be the equivalent channel from DU_i to DR, and $\mathbb{G} = \left[\mathfrak{g}_1 \cdots \mathfrak{g}_l \right]$, (3.4) can be written as

$$\mathbf{y}_r = P_d^{\frac{1}{2}} \mathfrak{g}_i t_i + \mathbb{G}_{-i}(\mathbf{t}_{-i})^T + \hat{\mathbf{H}} P_c^{\frac{1}{2}} \mathbf{m} + \mathbf{z}_r \tag{3.11}$$

where $\mathbf{t} = \left[t_1 \cdots t_l \right]$, $\mathbf{m} = \left[m_1 \cdots m_s \right]^T$, and $\hat{\mathbf{H}} = \left[\mathbf{H}_{CU,1}^r \mathbf{u}_1 \cdots \mathbf{H}_{CU,l}^r \mathbf{u}_s \right]$. $\mathbb{G}_{-i}$ denotes the matrix $\mathbb{G}$ without the i-th column.

Then, $\mathbf{U}_r$ is designed to cancel the interference from other DUs, while treating the interference from cellular links as noise. Specifically, let $\mathbf{u}_j \in \mathbb{C}^{1 \times N_d}$ denote the j-th row of $\mathbf{U}_r$, i.e., $\mathbf{U}_r = \begin{bmatrix} \mathbf{u}_1^T & \cdots & \mathbf{u}_l^T \end{bmatrix}^T$. $\mathbf{u}_j$ is designed as the null-space of $\mathbb{G}_{-j} \in \mathbb{C}^{N_d \times (l-1)}$, i.e.,

$$\mathbf{u}_j \cdot \mathbb{G}_{-j} = \mathbf{0} \tag{3.12}$$

Since $l \leq N_d$, we have $l - 1 < N_d$ for sure, which means $\mathbf{u}_j$ can be found for sure. As a result, the received signal of DU$_j$ can be expressed as

$$y_j = \omega_j P_d^{\frac{1}{2}} t_j + P_c^{\frac{1}{2}} \mathbf{u}_j \hat{\mathbf{H}} \mathbf{m} + \mathbf{u}_j \mathbf{z}_r, \quad j = 1, \cdots, l \tag{3.13}$$

where $\omega_j = \mathbf{u}_j \mathfrak{g}_j$.

3.3.2 Performance of Cellular Network

In this subsection, we investigate the outage probability of each cellular link, which is defined as the probability of event that ρ_i is lower than a predetermined threshold α, where ρ_i denotes the signal-to-noise ratio (SNR) of the link of CU$_i$. Note that based on the proposed scheme, the interference signals from DUs are all aligned in the orthogonal space of cellular links, which means there is no interference at each cellular link.

Theorem 3.2. *In a cellular uplink network underlaying a D2D LAN, based on the proposed 'interference-free' IA scheme, the outage probability of each cellular link is*

$$P_{out} = Pr[\rho_i \leq \alpha] = 1 - \frac{1}{\frac{\alpha}{P_c} + 1} \tag{3.14}$$

Proof. Since the interference is completely nulled, the received signals on BS can be written as (3.2). Hence, the SNR of CU$_i$, ρ_i can be expressed as

$$\rho_i = \frac{P_c |\lambda_i|^2}{|\mathbf{z}'_e(i)|^2} = P_c \frac{|\mathbf{D}_e \mathbf{H}^e_{CU,i} \mathbf{u}_i(i)|^2}{|\mathbf{D}_e \mathbf{z}_e(i)|^2} \tag{3.15}$$

where $\mathbf{D}_e \mathbf{H}^e_{CU,i} \mathbf{u}_i(i)$ and $\mathbf{D}_e \mathbf{z}_e(i)$ denotes the i-th row of vector $\mathbf{D}_e \mathbf{H}^e_{CU,i} \mathbf{u}_i$ and $\mathbf{D}_e \mathbf{z}_e$, respectively.

Let $y_i = \mathbf{D}_e \mathbf{z}_e(i)$. Since both $\mathbf{D}_e$ and $\mathbf{u}_i$ are unitary (each of them is the null space of a matrix), λ_i and y_i follow the same distribution as the elements in $\mathbf{H}^e_{CU,i}$ and $\mathbf{z}_e$, respectively. Hence, both λ_i and y_i have complex normal distribution with zero mean and unit variance, which means both $|\lambda_i|^2$ and $|y_i|^2$ have exponential distributions [36], i.e., $|\lambda_i|^2 \sim \exp(1)$ and $|y_i|^2 \sim \exp(1)$. Therefore, the probability density function (pdf) of $Z = \frac{|\lambda_i|^2}{|y_i|^2}$ is

$$f_Z(z) = \frac{1}{(z+1)^2}, \quad \text{for } z \geq 0 \tag{3.16}$$

Accordingly, the pdf of $\rho_i = P_c \cdot Z$ is

$$f_{\rho_i}(z) = \frac{1}{P_c(\frac{z}{P_c}+1)^2}, \quad \text{for } z \geq 0 \tag{3.17}$$

Finally, the outage probability of CU_i can be calculated as

$$P_{out} = Pr[\rho_i \leq \alpha] = \int_0^\alpha f_{\rho_i}(z) d_z = 1 - \frac{1}{\frac{\alpha}{P_c}+1} \tag{3.18}$$

3.3.3 Performance of D2D LAN

Next, we examine the outage probability of each D2D link. Let ρ_j^D denote the SINR of the link of DU_j. The outage probability of the D2D link is $Pr[\rho_j^D \leq \beta]$.

Theorem 3.3. *In a cellular uplink network underlaying a D2D LAN, based on the proposed 'interference-free' IA scheme, the outage probability of each D2D link is*

$$P_{out} = Pr[\rho_j^D \leq \beta] = 1 - \frac{1}{(\frac{\beta P_c}{P_d}+1)^s} \tag{3.19}$$

Proof. Let $\mathbf{p}_j \in \mathbb{C}^{1 \times s} = \mathbf{u}_j \hat{\mathbf{H}}$. According to (3.13), the SINR at DR with a high SNR approximation can be expressed as

$$\rho_j^D = \frac{P_d |\omega_j|^2}{P_c \mathbf{p}_j \mathbf{p}_j^* + |\mathbf{z}_r'|^2} = \frac{P_d |\omega_j|^2}{P_c \sum_{k=1}^s |p_k|^2 + |\mathbf{z}_r'|^2}$$

$$\approx \frac{P_d |\omega_j|^2}{P_c \mathbf{p}_j \mathbf{p}_j^* + |\mathbf{z}_r'|^2} = \frac{P_d |\omega_j|^2}{P_c \sum_{k=1}^s |p_k|^2} \tag{3.20}$$

where p_k denotes the k-th element of $\mathbf{p}_j$.

Since ω_j and p_k all follow independent complex Gaussian distribution, based on the results in [9], the pdf of ρ_j^D can be expressed as

$$f_{\rho_j^D}(x) = \frac{P_c}{P_d} \frac{s}{(x\frac{P_c}{P_d}+1)^{(s+1)}}, \quad \text{for } x \geq 0 \tag{3.21}$$

The outage probability of the link of DU$_j$ is calculated as

$$P_{out} = Pr[\rho_j^D \leq \beta] = \int_0^\beta f_{\rho_j^D}(x)\mathrm{d}x = 1 - \frac{1}{(\frac{\beta P_c}{P_d} + 1)^s} \qquad (3.22)$$

3.3.4 Discussion

The main advantage of this scheme is that the interference from DUs is completely nulled at the BS, which means the network is no longer interference limited. Hence, both cellular and D2D links can achieve the desired performance by adjusting the power. As can be seen from (3.14), the outage probability of each link of CU is only related to the power of CU, P_c. Further, in D2D LAN, (3.19) implies that the outage probability of the link of DU will decrease if the power of DU is increased. Hence, the DUs can adjust the power P_d to obtain the required QoS without considering its impact on cellular links.

On the other hand, we should also note that this scheme may not be applicable if the value of $\frac{N \cdot r}{N - M_d}$ is small. For example, if $\frac{N \cdot r}{N - M_d} \leq 1$, then the D2D LAN will not be active according to *Theorem 3.1*.

3.4 Interference-Limiting D2D Communication Scheme

In this scheme, the DUs' signals are allowed to occupy some links of CUs. However, the peak interference power on each of the 'interfered' links must be under a certain threshold γ.

3.4.1 Design Process

Three steps are involved in the design process. In **Step 1**, the BS designs and broadcasts the post-processing matrix $\mathbf{D}_e$. In **Step 2**, each DU designs the precoding vectors $\mathbf{v}_i$ according to the received $\mathbf{D}_e$. Each CU also designs precoding vector $\mathbf{u}_i$ as described in Sect. 3.2. In **Step 3**, some DUs are scheduled for transmission according to some user selection criterion.

- **Step 1:** We simply design $\mathbf{D}_e$ as an antenna selection matrix i.e.,

$$\mathbf{D}_e = \begin{bmatrix} \mathbf{I}_{s \times s} & \mathbf{0}_{s \times (N-s)} \end{bmatrix} \qquad (3.23)$$

where $\mathbf{I}_{s \times s}$ denotes the $s \times s$ identity matrix and $\mathbf{0}_{s \times (N-s)}$ denotes the $s \times (N - s)$ matrix with all zeros. Then, the CU_i can design $\mathbf{u}_i$ accordingly to realize the parallel channels on the first s antennas of BS as shown in (3.2).

- **Step 2:** We describe the design of $\mathbf{v}_i$ for DU_i. From (3.3) we can see that the j-th row of $\mathbf{D}_e \mathbf{H}^e_{DU,i}$ represents interference from DU_i added on the j-th link of CUs. Since $\mathbf{v}_i \in \mathbb{C}^{(M_d \times 1)}$, it can be designed as the null space of at most $M_d - 1$ rows of $\mathbf{D}_e \mathbf{H}^e_{DU,i}$, so as to keep the corresponding links free from interference.

  Assume that we want the first $M_d - 1$ links to be interference free, $\mathbf{v}_i$ can be designed by satisfying

$$\mathbf{D}_e(M_d - 1)\mathbf{H}^e_{DU,i} \cdot \mathbf{v}_i = \mathbf{0} \tag{3.24}$$

where $\mathbf{D}_e(M_d - 1)$ denotes the first $M_d - 1$ rows of $\mathbf{D}_e$.

- **Step 3:** Since all DUs design their signals to avoid interfering the first $M_d - 1$ links of CUs, it is equivalent to aligning the interference on to the rest $s - M_d + 1$ links of CUs. The received signal from DU_i on BS can be written as

$$\mathbf{D}_e \mathbf{H}^e_{DU,i} \mathbf{v}_i P_d^{\frac{1}{2}} t_i$$

$$= \left[\mathbf{0}_{1 \times (M_d-1)} \ \vartheta^i_{M_d} \ \cdots \ \vartheta^i_s \right]^T P_d^{\frac{1}{2}} t_i \tag{3.25}$$

where ϑ^i_k is a complex number that represents the interference from DU_i on the k-th link of BS. Accordingly, the interference power generated on the k-th link of BS by DU_i can be calculated as $|\vartheta^i_k|^2 P_d$.

Let γ denote the interference constraint on each link of CU, which can be determined according to the requirement of cellular network. The DU_i will become a qualified DU if the following condition is satisfied,

$$|\vartheta^i_k|^2 P_d \leq \frac{\gamma}{N_d}, \ \forall k = M_d, \cdots, s \tag{3.26}$$

Then, DR will select $l \leq N_d$ qualified DUs to start transmission. When there are a large number of DUs, the active DUs can be selected based on their equivalent channels with the semi-orthogonal user selection (SUS) algorithm introduced in [37].

3.4.2 Performance of Cellular Network

We first examine the performance of cellular network. Note that with this scheme, the first $M_d - 1$ links of CUs are free from interference. Therefore, they have the same performance as 'interference-free' IA scheme. Next, we focus on those 'interfered' links of CUs. Let ϕ_k denote the SINR of the k-th links of CUs, where $k \geq M_d$, we have the following Theorem,

Theorem 3.4. *In a cellular uplink network underlaying a D2D LAN, based on the proposed 'interference-limiting' IA scheme, the outage probability of each . 'interfered' cellular link is*

$$P_{out} = Pr[\phi_k \leq \alpha]$$

$$= 1 - (1 + \frac{\alpha}{P_c})^{-1} \cdot (1 - e^{-\gamma'})^{-N_d} \cdot$$

$$(1 + \frac{P_d \alpha}{P_c})^{-N_d} (1 - e^{-\gamma'(\frac{P_d \alpha}{P_c}+1)})^{N_d} \tag{3.27}$$

where $\gamma' = \frac{\gamma}{P_d N_d}$.

The proof is given in Appendix A of this Chapter.

3.4.3 Performance of D2D LAN

Note that in this scheme, only those DUs who can satisfy (3.26) are qualified to start D2D transmission. Hence, we first examine the number of qualified DUs for a certain threshold of a cellular link.

Proposition 3.1. *For the D2D LAN with K DUs underlaying a cellular uplink, given the interference constraint γ at each link of CU, the number of qualified DUs is*

$$N_q = K \cdot p_q \approx K \cdot (1 - e^{-\gamma'})^{(s-M_d+1)} \tag{3.28}$$

where p_q denotes the probability of a DU being qualified.

Proof. The proof is in Appendix B of this Chapter.

As can be seen, the value of N_q is related to γ and $s - M_d$. First, it is obvious that when the interference threshold is increased, more DUs will be allowed to transmit. In addition, we can see that with less links of CUs in the network (smaller s), more DUs are qualified for transmission.

Next, we study the performance of D2D LAN. Let $g_i \in \mathbb{C}^{N_d \times 1} = \mathbf{H}^r_{DU,i}\mathbf{v}_i$ denote the equivalent channel from DU_i to DR, the received signals on DR can be expressed as

$$\mathbf{y}_r = \underbrace{[g_1 \cdots g_l]}_{\mathscr{G} \in \mathbb{C}^{N_d \times l}} P_d^{\frac{1}{2}}\mathbf{t} + \mathbf{q}_c + \mathbf{z}_r \tag{3.29}$$

where $\mathbf{t} = [t_1 \cdots t_l]^T$, $\mathbf{q}_c = \Sigma_{i=1}^s \mathbf{H}^r_{CU,i}\mathbf{u}_i P_c^{\frac{1}{2}} m_i$, and $\mathbf{z}_r$ denotes the noise on DR.

Since $l \leq N_d$, the signal $\mathbf{t}$ can be decoded with a zero-forcing filter $\mathscr{G}^{-1}$, which leads to

$$\mathbf{y}'_r = P_d^{\frac{1}{2}}\mathbf{t} + \mathscr{G}^{-1}(\mathbf{q}_c + \mathbf{z}_r) \tag{3.30}$$

Then, the rate of D2D LAN can be expressed as

$$R_r = \log_2 \det(\mathbf{I} + \frac{\mathscr{G}\mathscr{G}^*}{(\mathbf{q}_c + \mathbf{z}_r)(\mathbf{q}_c + \mathbf{z}_r)^*}) \tag{3.31}$$

Note that the equivalent model in (3.29), (3.30) and (3.31) is the same as that of in [38]. Hence, it can be proved in the same way as in [38] that the sum rate of D2D LAN scales as $N_d \log_2(1 + \log(N_q))$, which means the performance of the network can be improved by taking advantage of multiuser diversity.

3.4.4 Discussion

As we can see, in this scheme, the number of active DUs is limited by $l \leq \min\{N_d, N_q\}$. Hence, as long as $N_q \geq N_d$, the number of active DUs can reach maximum. In addition, if there are a large number of DUs, this scheme can take advantage of multiuser diversity to improve the performance of D2D LAN.

3.5 Simulation Results

In this section, we examine the simulation performance of the networks.

Figures 3.2 and 3.3 show the theoretical and simulation results of outage probabilities of each link of CU and DU with the use of 'interference-free' scheme, respectively, where the x-axis is the threshold of SINR in dB. We set $N = 5$, $N_d = 4$, and $M = M_d = 3$. Three CUs communicate with BS, i.e., $s = 3$. Meanwhile, four DUs communicate with DR, whose signals are aligned in the two-dimensional orthogonal space of links of CUs at BS. In addition, the transmitting power of CU is set to 10 dB or 15 dB and the power of DU is set to 20 dB or 25 dB. The theoretical results are obtained from (3.14) and (3.19) for cellular link and D2D link, respectively. We can see that the simulation results are matched almost perfectly with theoretical results. The outage probability of each link decreases when their transmitting power is increasing. Moreover, the outage probability of the link of CU remains the same for different transmitting power on DU. This is because the interference signals from DU are completely nulled at links of CUs.

Figure 3.4 shows the number of qualified DUs in the application of 'interference-limiting' scheme. Specifically, it depicts the growth of the number of qualified DUs with an increase of total DUs, under different threshold γ. In the simulation, we

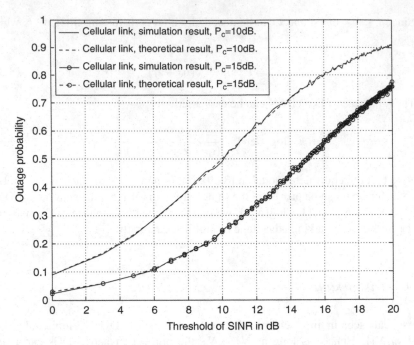

Fig. 3.2 Outage probability of each cellular link with 'interference-free' IA scheme. $N = 5$, $N_d = 4$, $M = M_d = 3$, $s = 3$, $P_c = 10\,\text{dB}$, $15\,\text{dB}$

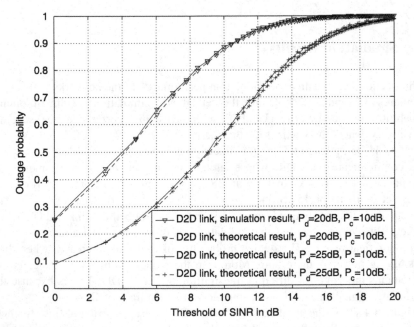

Fig. 3.3 Outage probability of each D2D link with 'interference-free' IA scheme. $N = 5$, $N_d = 4$, $M = M_d = 3$, $s = 3$, $P_c = 10\,\text{dB}$, $15\,\text{dB}$, $P_d = 20\,\text{dB}$, $25\,\text{dB}$

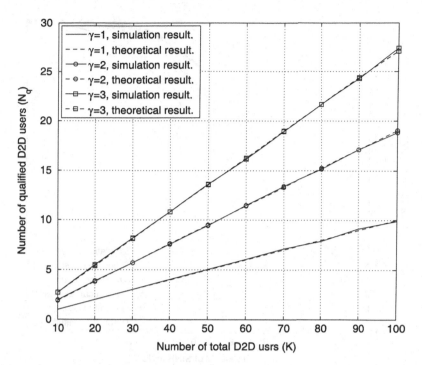

Fig. 3.4 Number of qualified D2D users with 'interference-limiting' IA scheme, $\gamma = 1, 2, 3, N_d = M = M_d = 3, N = 4$ and $s = 3, P_d = 5\,\text{dB}$

set three CUs communicating with BS, i.e., $s = 3$. In addition, each DU designs the precoding vector such that the interference signal does only affect the third link of CU. Then, the DU who can satisfy (3.26) is selected as a qualified user. The theoretical results are obtained from (3.28). As we can see, the theoretical results are matched with simulation results perfectly. Hence, *Proposition 3.1* is verified in the figure. For instance, the number of qualified users increases almost linearly with the increase of total number of DUs. In addition, higher threshold leads to more qualified DUs because the probability of a DU satisfying (3.26) becomes higher.

Figure 3.5 shows the simulation results and theoretical results of the outage probability of each interfered cellular link with 'interference-limiting' scheme, under different interference threshold γ. The system setting in this simulation is the same as that in Fig. 3.4. The theoretical result is obtained according to (3.27). As we can see, the outage probability of the interfered cellular link increases when the interference threshold increases. This is because more interference power is allowed on the cellular link.

Finally, in Fig. 3.6 we study the sum rate of the D2D LAN achieved with 'interference-limiting' schemes, under different transmitting power of DU and different number of DUs. We set $M = M_d = 3$ and $N = N_d = 4$. Three CUs communicate with the BS, i.e., $s = 3$. In 'interference-limiting' scheme, SUS

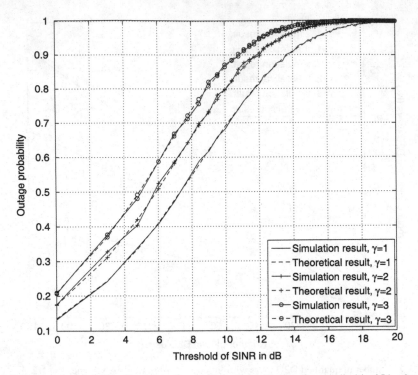

Fig. 3.5 Outage probability of each interfered cellular link with 'interference-limiting' IA scheme, $\gamma = 1, 2, 3, N_d = M = M_d = 3, N = 4$ and $s = 3, P_c = 10\,\mathrm{dB}, P_d = 5\,\mathrm{dB}$

algorithm [37] is used to select active DUs from qualified DUs. Hence, the sum rate of D2D LAN can be calculated according to (3.31). Note that at most four DUs can be selected. As can be seen, 'interference-limiting' scheme can take advantage of multiuser diversity. When there are a large number of DUs, the performance of D2D LAN can be largely improved.

3.6 Summary

In this Chapter, we investigated interference coordination techniques in D2D LAN underlaying a cellular uplink, where multiple DUs intend to communicate with a D2D receiver. Two schemes were proposed to effectively manage the mutual interference between the two networks for different scenarios. 'Interference-free' scheme is applicable when the number of DUs is small and the cellular links are not fully occupied, whereas 'Interference-limiting' scheme can be used in other cases. The performance of cellular and D2D networks with both schemes were analyzed. The theoretical results were corroborated by simulations.

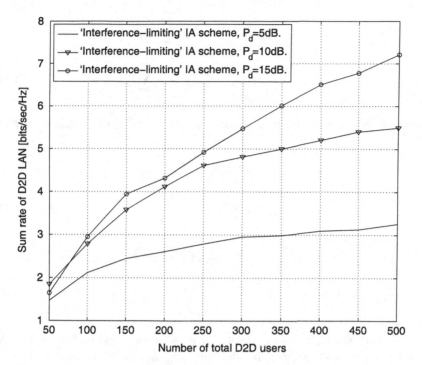

Fig. 3.6 Achievable sum rate of D2D LAN with 'interference-limiting' schemes, $\gamma = 2$, $M = M_d = 3$, $N = N_d = 4$ and $s = 3$, $P_c = 5\,\text{dB}$, $P_d = 5\,\text{dB}$, $10\,\text{dB}$, $15\,\text{dB}$

Appendix A: Proof of Theorem 3.4

The SINR of the k-th link of CUs can be expressed as

$$\phi_k = \frac{P_c |\lambda_k|^2}{|\mathbf{z}'_e(k)|^2 + P_d \sum_{i=1}^{N_d} |\vartheta_k^i|^2} \tag{3.32}$$

First, we let Λ, Z, and Θ_i denote $|\lambda_k|^2$, $|\mathbf{z}'_e(k)|^2$ and $|\vartheta_k^i|^2$, respectively, where their PDF is regardless of index k. In other words, they are i.i.d. for all k. Hence, (3.32) can be written as

$$\phi_k = \frac{P_c \Lambda}{Z + P_d \sum_{i=1}^{N_d} \Theta_i} \tag{3.33}$$

Accordingly, the outage probability of the k-th 'interfered' link of CUs can be expressed as follows,

$$P_{out} = Pr[\phi_k \leq \alpha]$$

$$= Pr[\Lambda \le \frac{\alpha}{P_c}(Z + P_d \sum_{i=1}^{N_d} \Theta_i)]$$

$$= 1 - Pr[\Lambda > \frac{\alpha}{P_c}(Z + P_d \sum_{i=1}^{N_d} \Theta_i)] \tag{3.34}$$

Since Λ has exponential distribution, i.e., $\Lambda \sim \exp(1)$, (3.34) can be further written as

$$P_{out} = 1 - \mathrm{E}\{e^{-\frac{\alpha}{P_c}(Z + P_d \sum_{i=1}^{N_d} \Theta_i)}\}$$

$$= 1 - \mathrm{E}\{e^{-\frac{\alpha}{P_c}Z}\} \cdot \prod_{i=1}^{N_d} \mathrm{E}\{e^{-\frac{\alpha P_d}{P_c}\Theta_i}\} \tag{3.35}$$

$$= 1 - \int_0^{+\infty} e^{-\frac{\alpha}{P_c}z} f_Z(z) dz \cdot \prod_{i=1}^{N_d} \int_0^{\gamma'} e^{-\frac{\alpha P_d}{P_c}\theta} f_{\Theta_i}(\theta) d\theta$$

$$\tag{3.36}$$

where $\gamma' = \frac{\gamma}{P_d N_d}$.

Equation (3.35) is due to the fact that the variables Z and Θ_i for $i = 1, \cdots N_d$ are independent of each other. In addition, $f_Z(z)$ and $f_{\Theta_i}(\theta)$ denote the PDF of variables Z and Θ_i, respectively. Based on **Theorem 3.2**, it can be shown that Z has exponential distribution, i.e., $Z \sim \exp(1)$, which means

$$f_Z(z) = e^{-z} \quad \text{for} \quad z \ge 0 \tag{3.37}$$

Next, we derive $f_{\Theta_i}(\theta)$. Since Θ_i are i.i.d for $i = 1, \cdots, N_d$, the index i of Θ_i is omitted in the following discussions. Note that Θ is from 0 to γ', the CDF of Θ is

$$F_\Theta(\theta) = \frac{1}{1 - e^{-\gamma'}}(1 - e^{-\theta}) \quad \text{for} \quad 0 \le \theta \le \gamma' \tag{3.38}$$

which leads to

$$f_\Theta(\theta) = F'_\Theta(\theta) = \frac{e^{-\theta}}{1 - e^{-\gamma'}} \tag{3.39}$$

Then, by taking (3.37) and (3.39) into (3.36), we can obtain (3.27).

Appendix B: Proof of Proposition 3.1

The interference constraint on each link is γ, which can be surely guaranteed by limiting the interference generated by each DU to be less than $\frac{\gamma}{l}$. Further, since $l \leq N_d$, we have (3.26) as the condition that should be satisfied for each selected DU, which is equivalent to

$$|\vartheta_{k'}^i|^2 \leq \gamma', \ \forall k' = M_d, M_d + 1, \cdots, s \qquad (3.40)$$

where $\gamma' = \frac{\gamma}{P_d N_d}$.

Hence, the DU who can satisfy (3.40) is qualified for D2D communications with DR. Let p_i denote the probability of DU_i satisfying (3.40), we have $p_1 = p_2 = \cdots = p_K = p_q$, which is equal to the probability of $|\vartheta_{k'}^i|^2 \leq \gamma', \ \forall k' = M_d, M_d + 1, \cdots, s$.

Further, since the CDF of $|\vartheta_{k'}^i|^2$ is given in (3.38) as $F_\vartheta(x)$, we have

$$p_q = [F_\vartheta(\gamma')]^{s - M_d + 1} = (1 - e^{-\gamma'})^{s - M_d + 1} \qquad (3.41)$$

Finally, the average number of qualified users satisfying the interference constraint is $N_q \approx K(1 - e^{-\gamma'})^{s - M_d + 1}$.

References

1. Doppler, K., Rinne, M., Wijting, C., Ribeiro, C. B., & Hugl, K. (2009). Device-to-device communication as an underlay to LTE-advanced networks. *IEEE Communications Magazine, 47*(12), 42–49.
2. Lin, X., Andrews, J., Ghosh, A., & Ratasuk, R. (2014). An overview of 3GPP device-to-device proximity services. *IEEE Communications Magazine, 52*(4), 40–48.
3. Doumi, T., Dolan, M., Tatesh, S., Casati, A., Tsirtsis, G., Anchan, K., & Flore, D. (2013). LTE for public safety networks. *IEEE Communications Magazine, 51*(2), 106–112.
4. Tang, H., Ding, Z., & Levy, B. (2014). Enabling D2D communications through neighbor discovery in LTE cellular neworks. *IEEE Transactions on Signal Processing, 62*(19), 5157–5170.
5. Fodor, G., Dahlman, E., Mildh, G., Parkvall, S., Reider, N., Miklos, G., & Turanyi, Z. (2012). Design aspects of network assisted device-to-device communications. *IEEE Communications Magazine, 50*(3), 170–177.
6. Lin, X., Andrews, J., & Ghosh, A. (2014). Spectrum sharing for device-to-device communication in cellular networks. *IEEE Transactions on Wireless Communications, 13*(12), 6727–6740.
7. Asadi, A., Wang, Q., & Mancuso, V. (2014). A survey on device-to-device communication in cellular networks. *IEEE Communications Surveys and Tutorials, 16*(4), 1801–1819. Fourth Quarter
8. Yu, C.-H., Doppler, K., Ribeiro, C. B., & Tirkkonen, O. (2011). Resource sharing optimization for device-to-device communication underlaying cellular networks. *IEEE Transactions on Wireless Communications, 10*(8), 2752–2763.
9. Min, H., Lee, J., Park, S., & Hong, D. (2011). Capacity enhancement using an interference limited area for device-to-device uplink underlaying cellular networks. *IEEE Transactions on Wireless Communications, 10*(12), 3995–4000.

10. Feng, D., Lu, L., Wu, Y., Li, G. Y., Feng, G., & Li, S. (2013). Device-to-device communications underlaying cellular networks. *IEEE Transactions on Communications, 61*(8), 3541–3551.
11. Zhu, D., Wang, J., Swindlehurst, A. L., & Zhao, C. (2014). Downlink resource reuse for device-to-device communications underlaying cellular networks. *IEEE Signal Processing Letters, 21*(5), 531–534.
12. Lee, D., Choi, K., Jeon, W., & Jeong, D. (2014). Two-stage semi-distributed resource management for device-to-device communication in cellular networks. *IEEE Transactions on Wireless Communications, 13*(4), 1908–1920.
13. Lee, N., Lin, X., Andrews, J., & Heath, R., Jr. (2015). Power control for D2D underlaid cellular networks: Modeling, algorithms, and analysis. *IEEE Journal on Selected Areas in Communications, 33*(1), 1–13.
14. Xu, C., Song, L., Han, Z., Zhao, Q., Wang, X., Cheng, X., & Jiao, B. (2013). Efficient resource allocation for device-to-device underlaying communication systems: A reverse iterative combinatorial auction based approach. *IEEE Journal on Selected Areas in Communications, 31*(9), 348–358.
15. Song, L., Niyato, D., Han, Z., & Hossain, E. (2014). Game-theoretic resource allocation methods for device-to-device (D2D) communication. *IEEE Wireless Communications Magazine, 21*(3), 136–144.
16. Boccardi, F., Clerckx, B., Ghosh, A., Hardouin, E., Jongren, G., Kusume, K., Onggosanusi, E., & Tang, Y. (2012). Multiple antenna techniques in LTE-advanced. *IEEE Communications Magazine, 50*(3), 114–121.
17. Lim, C., Yoo, T., Clerckx, B., Lee, B., & Shim, B. (2013). Recent trend of multiuser MIMO in LTE-Advanced. *IEEE Communications Magazine, 51*(3), 127–135.
18. Zhang, R., & Cui, S. (2010). Cooperative interference management with MISO beamforming. *IEEE Transactions on Signal Processing, 58*(10), 5450–5458.
19. Shaverdian, A., & Nakhai, M. (2014). Robust distributed beamforming with interference coordination in downlink cellular networks. *IEEE Transactions on Communications, 62*(7), 2411–2421.
20. Yang, L., & Zhang, W. (2015). On degrees of freedom region of three-user MIMO interference channels. *IEEE Transactions on Signal Processing, 63*(3), 590–603.
21. Phunchongharn, P., Hossain, E., & Kim, D. I. (2013). Resource allocation for device-to-device communications underlaying LTE-advanced networks. *IEEE Wireless Communications Magazine, 20*(4), 91–100.
22. Janis, P., Koivunen, V., Ribeiro, C., Doppler, K., & Hugl, K. (2009). Interference-avoiding MIMO schemes for device-to-device radio underlaying cellular network. In *Proceedings of IEEE Personal, Indoor and Mobile Radio Communications (PIMRC)*, Tokyo, pp. 2385–2389, 13–16 Sept 2009.
23. Xu, W., Liang, L., Zhang, H., Jin, S., Li, J., & Lei, M. (2012). Performance enhanced transmission in device-to-device communications: Beamforming or interference cancellation? In *Proceedings of IEEE Global Communications Conference (GLOBECOM 2012)*, Anaheim, pp. 4296–4301, 3–7 Dec 2012.
24. Tang, H., Zhu, C., & Ding, Z. (2013). Cooperative MIMO precoding for D2D underlay in cellular networks. In *Proceedings of IEEE International Conference on Communications (ICC)*, Budapest, pp. 5517–5521, 9–13 June 2013.
25. Jayasinghe, K., Jayasinghe, P., Rajatheva, N., & Latva-Aho, M. (2014). Linear precoder-decoder design of MIMO device-to-device communication underlaying cellular communication. *IEEE Transactions on Communications, 62*(12), 4303–4319.
26. Elkotby, H., Elsayed, K., & Ismail, M. H. (2012). Exploiting interference alignment for sum rate enhancement in D2D-enabled cellular networks. In *Proceedings of IEEE Wireless Communications and Networking Conference (WCNC)*, Paris, pp. 1624–1629, 1–4 Apr 2012.

27. Jiang, J., Peng, M., Wang, W., & Zhang, K. (2013). Energy efficiency optimization based on interference alignment for device-to-device MIMO downlink underlaying cellular network. In *Proceedings of IEEE Global Communications Conference (Globecom 2013)-International Workshop on Device-to-Device (D2D) Communication With and Without Infrastructure*, Atlanta, pp. 585–590, 9–13 Dec 2013.
28. Lin, X., Heath, R., Jr., & Andrews, J. (2015). The interplay between massive MIMO and underlaid D2D networking. *IEEE Transactions on Wireless Communications, 14*(6), 3337–3351.
29. Cho, S., Huang, K., Kim, D., Lau, V., Chae, H., Seo, H., & Kim, B. (2012). Feedback-topology designs for interference alignment in MIMO interference channels. *IEEE Transactions on Signal Processing, 60*(12), 6561–6575.
30. Yang, L., & Zhang, W. (2013). Opportunistic interference alignment in heterogeneous two-cell uplink network. In *Proceedings of IEEE International Conference on Communications (ICC 2013)*, Budapest, pp. 5448–5452, 9–13 June 2013.
31. Ayach, O., & Heath, R., Jr. (2012). Interference alignment with analog channel state feedback. *IEEE Transactions on Wireless Communications, 11*(2), 626–636.
32. Ayach, O., Lozano, A., & Heath, R., Jr. (2012). On the overhead of interference alignment: Training, feedback, and cooperation. *IEEE Transactions on Wireless Communications, 11*(11), 4192–4203.
33. Jindal, N. (2006). MIMO broadcast channels with finite-rate feedback. *IEEE Transactions on Information Theory, 52*(11), 5045–5060.
34. Love, D., Heath, R., Jr., Lau, V. K. N., Gesbert, D., Rao, B. D., & Andrews, M. (2008). An overview of limited feedback in wireless communication systems. *IEEE Journal on Selected Areas in Communications, 26*(8), 1341–1365.
35. Cho, S., Huang, K., Kim, D., Lau, V. K. N., Chae, H., Seo, H., & Kim, B. (2012). Feedback-topology designs for interference alignment in MIMO interference channels. *IEEE Transactions on Signal Processing, 60*(12), 6561–6575.
36. Gradshteyn, I. S., & Ryzhik, I. M. (1994). *Tables of integrals, series, and products*. San Diego: Academic.
37. Yoo, T. & Goldsmith, A. (2006). On the optimality of multiantenna broadcast scheduling using zero-forcing beamforming. *IEEE Journal on Selected Areas in Communications, 24*(3), 528–541.
38. Yang, L., Zhang, W., Zheng, N., & Ching, P. C. (2014). Opportunistic user scheduling in MIMO cognitive radio networks. In *Proceedings of IEEE International Conference on Acoustics, Speech and Signal Processing (ICASSP 2014)*, Florence, pp. 7303–7307, 4–9 May 2014.

Chapter 4
Interference Coordination in Cognitive Radio Systems

4.1 Introduction

Heterogeneous cellular network is a wireless communications paradigm that will be adopted by 5G network, where multiple communication systems coexist in the same network. This feature advocates the dense deployment of access points and extensive reuse of spectrum resources, which enables the network to accommodate the exponentially growing mobile data traffic. In addition, in order to well integrate different systems, the 5G network should be a "cognitive network" such that different systems can be aware of each other [1]. Hence, in 5G network, two communication systems may often be thought of as a cognitive radio (CR) network where one system is the primary system and the other one is secondary system, e.g. D2D communication systems and cellular systems [2]; macrocell system and femtocell system [3], etc.

Cognitive radios are software defined radios which are aware of their environment, and have the ability to learn from and quickly adapt to variations of their environment and the network requirements by changing their transmission parameters [4]. It allows a class of radio devices, called secondary users (SUs), to opportunistically access certain portions of spectrum, called white spaces, that are not occupied by licensed users [5]. Conventionally, the white spaces are usually associated with time and/or frequency domain, which appear mainly when either transmissions in the primary network are sporadic, i.e., there are time or frequency slots over which no transmission takes place, or there is no network infrastructure for the primary system in a given area [6]. However, in the case of dense networks,

© The Author(s) 2015
L. Yang, W. Zhang, *Interference Coordination for 5G Cellular Networks*,
SpringerBriefs in Electrical and Computer Engineering,
DOI 10.1007/978-3-319-24723-6_4

e.g. 5G cellular network, white space can be a rare and/or short lasting event. Consequently, in the absence of such spectrum holes, secondary systems are unable to transmit without producing additional interference to the primary systems.

One effective solution to this situation is to exploit the white space in spatial domain, by taking advantage of MIMO techniques. As mentioned in Chap. 2, multiple antennas can be utilized to achieve many desirable functions for wireless communications [7, 8]. In the context of CR, multi-antennas can be used to allocate transmit dimensions in space and hence provide the secondary transmitters more degrees of freedom in space in addition to time and frequency so as to coordinate the interference powers at the primary receivers [9].

In this chapter, we introduce interference coordination techniques for two representative cognitive radio networks, which are cognitive interference network, and cognitive uplink network. For interference network, we discuss how to explore the free spatial resources of primary links by implementing multi-antenna beamforming at the secondary transmitter. An interference coordination is introduced to eliminate the interference on primary system. Then, a cognitive uplink network is considered. In addition to MIMO techniques, we further improve the performance of secondary system by taking advantage of multiuser diversity.

The rest of the Chapter is organized as follows. In Sect. 4.2, the interference coordination techniques in cognitive interference network is studied. In Sect. 4.3, the interference coordination techniques in cognitive uplink network is studied. Section 4.4 summarizes the Chapter.

4.2 Cognitive Interference Network

In this section, we consider cognitive interference network, where two unidirectional links simultaneously operating in the same frequency band and producing mutual interference, as shown in Fig. 4.1. The primary system consists of a primary transmitter (PT) and a primary receiver (PR), where the PT and PR are equipped with M_p and N_p antennas, respectively. The secondary system consists of a secondary transmitter (ST) and a secondary receiver (SR), where the ST and SR are equipped with M_s and N_s antennas, respectively. The primary system is unaware of the existence of the secondary system, and must operate free of any additional interference produced by secondary systems.

In order to maximize the transmission rate, a water-filling power allocation scheme is implemented at the primary system over the spatial directions associated with the singular values of its channel matrix [10]. Interestingly, even if the PT maximizes the transmission rate, some of the spatial directions (SD) are usually left unused due to power limitations. The unused SD can therefore be reused by another (secondary) system operating in the same frequency band. Specifically, an ST can send its own data to its respective receiver by processing its signal in such a way that the interference produced on the primary link impairs only the unused SDs [6].

Next, we explain the interference coordination technique that is based on the idea described above.

Fig. 4.1 Cognitive
interference network

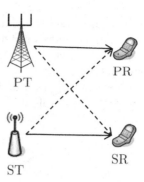

PT

PR

((•))

ST

SR

4.2.1   Interference Coordination

We first present the signal transmission of primary link. Without secondary link, the
received signals on PR can be expressed as

$$\mathbf{y}_p = \mathbf{D}_p \mathbf{H}_{pp} \mathbf{V}_p \mathbf{P}^{\frac{1}{2}} \mathbf{m}_p + \mathbf{D}_p \mathbf{z}_p \tag{4.1}$$

where $\mathbf{H}_{pp} \in \mathbb{C}^{N_p \times M_p}$ denotes the channel between PT and PR, and $\mathbf{V}_p \in \mathbb{C}^{M_p \times M_p}$
and $\mathbf{D}_p \in \mathbb{C}^{N_p \times N_p}$ are the precoding matrix on PT and post-processing matrix on
PR, respectively. $\mathbf{m}_p \in \mathbb{C}^{M_p \times 1}$ denotes the original message vector sent from PT,
in which the first s elements are the symbols to be transmitted and the last $M_p - s$
elements are zeros. The value of s equals the number of non-zero elements in the
main diagonal of $\mathbf{P}$, where $\mathbf{P} = \mathrm{diag}\left(p_1 \cdots p_{M_p}\right)$ denotes the power allocation
matrix on PT and $\mathbf{z}_p$ denotes the noise with variance $\sigma_p^2 \mathbf{I}_{N_p}$.

Let $\mathbf{H}_{pp} = \mathbf{U}_{pp} \Lambda_{pp} \mathbf{V}_{pp}^H$ be a singular value decomposition (SVD) of $\mathbf{H}_{pp}$, where
$\mathbf{U}_{pp} \in \mathbb{C}^{N_p \times N_p}$ and $\mathbf{V}_{pp} \in \mathbb{C}^{M_p \times M_p}$ are two unitary matrices, and $\Lambda_{pp} \in \mathbb{C}^{N_p \times M_p}$ with
main diagonal $\left(\lambda_{p,1} \cdots \lambda_{p,\min\{N_p,M_p\}}\right)$ and zeros on its off-diagonal. Using water-
filling power allocation, the primary link maximizes capacity by choosing $\mathbf{V}_p = \mathbf{V}_{pp}$
and $\mathbf{D}_p = \mathbf{U}_{pp}^H$, $\mathbf{P} = \mathrm{diag}\left(p_1 \cdots p_{M_p}\right)$ with

$$p_k = (a - \frac{\sigma_p^2}{\lambda_{p,k}^2})^+, \quad k = 1,2,\cdots M_p \tag{4.2}$$

where $(A)^+$ denotes $\max\{A, 0\}$, and the constant a (water-level) is set to satisfy the
power constraint. If $p_k > 0$, it means the kth sub-channel is used by the primary link
(known as used SDs). Accordingly, (4.1) can be written as

$$\mathbf{y}_p = \Lambda_{pp} \mathbf{P}^{\frac{1}{2}} \mathbf{m}_p + \mathbf{D}_p \mathbf{z}_p \tag{4.3}$$

where the equivalent channel $\Lambda_{pp}\mathbf{P}^{\frac{1}{2}} \in \mathbb{C}^{N_p \times M_p}$ is a diagonal matrix whose main diagonal contains s nonzero entries and $n = N_p - s$ zero entries. The value of s is equal to the number of SDs that are used by the primary link and n is the number of unused SDs.

Next, we discuss the design of transmit signals of ST. With secondary system, the received signals at PR become

$$\mathbf{y}_p = \Lambda_{pp}\mathbf{P}^{\frac{1}{2}}\mathbf{m}_p + \underbrace{\mathbf{D}_p\mathbf{H}_{pr}\mathbf{v}_r\mathbf{P}_r^{\frac{1}{2}}m}_{SU\text{interference}} + \mathbf{D}_p\mathbf{z}_p \tag{4.4}$$

where $\mathbf{H}_{pr}$ denotes the channel from ST to PR, $\mathbf{v}_r$, m, and $\mathbf{P}_r$ denote the precoding vector, original message, and transmitting power of ST, respectively.

Note that the first s rows of $\mathbf{D}_p$ correspond to the sub-channels that are used by the primary link, which are not supposed to be interfered. Hence, the power of interference that is caused by the secondary system on the used sub-channels can be expressed as

$$|\mathbf{I}|^2 = |\mathbf{D}_p(s)\mathbf{H}_{pr}\mathbf{v}_r|^2 \cdot \mathbf{P}_r \tag{4.5}$$

where $\mathbf{D}_p(s) \in \mathbb{C}^{s \times N_p}$ denotes the first s rows of $\mathbf{D}_p$, $|\mathbf{A}|$ denotes the L_2-norm of vector $\mathbf{A}$.

Therefore, to ensure that the primary system is not affected by the secondary system, $|\mathbf{I}|^2 = \mathbf{0}$ must be hold, which is equivalent to

$$\mathbf{G}\mathbf{v}_r = \mathbf{0} \tag{4.6}$$

where $\mathbf{G} \in \mathbb{C}^{s \times M_s} = \mathbf{D}_p(s)\mathbf{H}_{pr}$.

As can be seen, the non-zero solution of $\mathbf{v}_r$ in (4.6) exists only when $s < M_s$. This implies that to completely eliminate the impact of interference on PR, the number of antennas at ST must be larger than the number of SD being used by primary system.

The detailed explanation of the techniques described above can be found in [6, 11, 12].

4.3   Cognitive Uplink Network

In Sect. 4.2, we discussed the interference coordination scheme for cognitive interference networks. There are two problems remaining unsolved. First, if the number of antennas at ST is not larger than the number of SD used by primary system, i.e., $s \geq M_s$, how can the secondary system still be active? Second, while the performance of the primary system is guaranteed, how to improve the transmit rate of secondary system.

In this section, we address these problems by studying the scenario where the secondary system is a multi-user uplink MIMO network. (The primary system is still a MIMO point-to-point system.) To tackle the first problem, we have to 'relax' the interference constraint of primary system, i.e., interference is allowed at each link of primary system, but must be under a pre-determined threshold. Then, in order to improve the performance of secondary system, we take advantage of multi-user diversity, i.e., when there is a large number of secondary users (SUs) in presence, some SUs can be opportunistically selected to obtain the optimal performance [13–17]. Specifically, a two-stage opportunistic user scheduling scheme is proposed in this chapter, which enables the secondary link to take advantage of multiuser diversity while ensuring that the interference of primary link is under a certain threshold. We first select some SUs that cause the interference at primary receiver (PR) less than a predetermined threshold. These pre-selected SUs are referred to as candidate users. In the second stage, N_s users are selected from the candidate users according to semi-orthogonal user selection (SUS) algorithm [18, 19] to transmit signals to SR, where N_s denotes the number of antennas on SR. It is shown that with large K, (the total number of SUs) the sum rate of secondary link scales as $N_s \log \log K$, i.e., the multiuser diversity gain can be obtained. Meanwhile, the interference caused on the primary link is constrained below a preset threshold.

4.3.1   System Model

We first briefly introduce the system model. As shown in Fig. 4.2, the secondary system is now a multi-user uplink, where multiple secondary users (SUs) intend to communicate with PT. The antenna configurations on PT, PR, and SR remain to be the same as Sect. 4.2. Further, we assume each SU is equipped with M_s antennas.

The signal transmission of the primary system is exactly the same as (4.1), (4.2) and (4.3). In secondary system, the SUs are opportunistically selected to transmit signals by circumventing the SDs being used by the primary link. We let $\mathbf{H}_{pi} \in \mathbb{C}^{N_p \times M_s}$ and $\mathbf{H}_{si} \in \mathbb{C}^{N_s \times M_s}$ denote the channel between SU i and PR and the channel between SU i and SR, respectively.

Hence, similar to (4.4) and (4.5), the power of interference that is caused by SU_i on the used sub-channels can be expressed as

$$|\mathbf{I}_i|^2 = |\mathbf{D}_p(s)\mathbf{H}_{pi}\mathbf{v}_i|^2 \cdot \mathbf{P}_i \qquad (4.7)$$

where $\mathbf{v}_i$ and $\mathbf{P}_i$ denote the precoding vector and transmit power on SU_i, respectively.

Next, we discuss the design of $\mathbf{v}_i$ and the selection of active SUs.

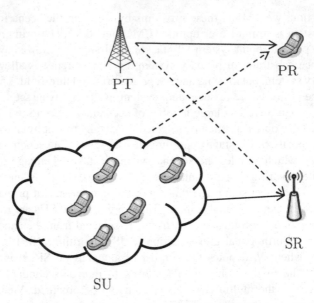

Fig. 4.2 The cognitive radio network where the primary link is a point-to-point MIMO channel and the secondary link is a multi-user MIMO uplink with K secondary users

4.3.2 Interference Coordination Techniques

In this section, we propose a two-stage secondary user scheduling scheme. In the first stage, the SUs who may cause interference on PR less than a certain threshold are selected, which are referred to as the candidate users. In the second stage, N_s candidate users are further selected according to a semi-orthogonal user selection (SUS) [18, 19]-based algorithm to transmit signals.

4.3.3 Stage 1: Selection of Candidate Users

We first introduce the design of precoding vector of each SU. Then, the criteria of candidate users is given.

Let $\mathbf{G}_i \in \mathbb{C}^{s \times M_s} = \mathbf{D}_p(s)\mathbf{H}_{pi}$. As mentioned before, if $s < M_s$, $\mathbf{G}_i$ is a 'fat' matrix, which means $\mathbf{v}_i$ can be set as the null space of $\mathbf{G}_i$ to ensure that $\mathbf{G}_i\mathbf{v}_i = 0$, i.e., the interference on the "in use" sub-channels are zero. However, if $s \geq M_s$, $\mathbf{G}_i$ becomes a square or 'tall' matrix, which makes $\mathbf{G}_i\mathbf{v}_i = 0$ impossible almost for sure. In that case, $\mathbf{v}_i$ needs to be designed such that $|\mathbf{G}_i\mathbf{v}_i|^2$ is minimized. Let $\mathbf{G}_i = \Omega_i \Sigma_i \Phi_i^H$ denote the SVD of $\mathbf{G}_i$, where $\Omega_i \in \mathbb{C}^{s \times s}$ and $\Phi_i \in \mathbb{C}^{M_s \times M_s}$ consist of orthonormal columns, and $\Sigma_i \in \mathbb{C}^{s \times M_s}$ has main diagonal $(\lambda_{i,1}, \cdots, \lambda_{i,M_s}, 0, \cdots, 0)$ and zeros on off-diagonal. In addition, $\lambda_{i,1} \geq \lambda_{i,2} \geq \cdots \geq \lambda_{i,\min\{s,M_s\}}$. Then, $\mathbf{v}_i$ should be set as the last column of Φ_i, such that

$$|\mathbf{I}_i|^2 = |\mathbf{G}_i\mathbf{v}_i|^2\mathbf{P}_i = \lambda_{i,M_s}^2|\mathbf{v}_i|^2\mathbf{P}_i = \lambda_{i,M_s}^2\mathbf{P}_i \qquad (4.8)$$

$\lambda_{i,M_s}^2\mathbf{P}_i$ would be the smallest interference power that SU i can have, which is referred to as the "interference leakage" of SU i.

As we can see, in the case of $s \geq M_s$, interference on PR is inevitable with the presence of secondary link. To protect the performance of primary link while improving the spectrum efficiency of the network, we let the total interference power on PR be lower than a predetermined threshold, γ_{th},

$$\sum_{i=1}^{l}|\mathbf{I}_i|^2 = \sum_{i=1}^{l}\lambda_{i,M_s}^2\mathbf{P}_i \leq \gamma_{th} \qquad (4.9)$$

(4.9) can be guaranteed by setting

$$\lambda_{i,M_s}^2 \leq \frac{\gamma_{th}}{\mathbf{P}_iN_s} \triangleq \gamma_{th}' \qquad (4.10)$$

Hence, SU i would be deemed as a candidate user if (4.10) can be satisfied.

4.3.4 Stage 2: Selection of Active Users

Assume N_c candidate users are selected in stage 1. Then, further N_s of them are selected to transmit signals to SR according to a selection algorithm which will be explained later. The received signals on SR can be expressed as

$$\mathbf{y}_r = \underbrace{\left[\mathbf{H}_{s1}\mathbf{v}_1 \cdots \mathbf{H}_{sl}\mathbf{v}_l\right]}_{\mathcal{H}}\mathbf{P}_i^{\frac{1}{2}}\mathbf{m}_s + \mathbf{I}_p + \mathbf{z}_s$$

$$= \left[\mathfrak{h}_1 \cdots \mathfrak{h}_l\right]\mathbf{P}_i^{\frac{1}{2}}\mathbf{m}_s + \mathbf{I}_p + \mathbf{z}_s \qquad (4.11)$$

where $\mathbf{I}_p$ denotes the interference term from primary link, $\mathbf{z}$ denotes the noise on SR. Since $\mathbf{v}_i$ has already been designed in the first stage, $\mathfrak{h}_i \in \mathbb{C}^{N_s \times 1} = \mathbf{H}_{si}\mathbf{v}_i$, ($i = 1, 2, \cdots l$) can be seen as the equivalent channel between SU i and SR.

Next, we explain how to select these l candidate users. As $K >> N_s$, for large K it has $N_c > N_s$ almost surely. Then, N_s candidate users are selected finally, based on a SUS-based algorithm, which is shown in **Algorithm 1**.

The algorithm can be explained as follows. In **step 2**, the user who has the largest equivalent channel gain $|\mathfrak{h}_i|^2$ is selected and the first basis vector $\mathbf{g}_{(1)}$ is set accordingly. In **step 3**, we apply the SUS [18] to select other candidate users. Since **step 3** is a loop, we first explain **step 3**-(b). In (b), we project equivalent user channels in Q_i to the orthogonal complement of span$\{\mathbf{g}_{(1)}, \cdots, \mathbf{g}_{(i-1)}\}$. Next in (c), the user with the largest projected norm is selected as $S(i)$, which determines

---

Algorithm 1 Semi-orthogonal user selection

Step 1. Initialization, $S = \varnothing$, $i = 1$.
The candidate SUs are denoted by
$Q_1 = \{k \mid \lambda_{k,M_s}^2 \leq \gamma_{th}, k = 1, \cdots, N_c\}$,
Step 2. Selecting the first user
1) The first selected SU is determined by
$S(1) = \arg\max_{k \in Q_1} |\mathfrak{h}_k|^2$, where $\mathfrak{h}_k \triangleq \mathbf{H}_{sk}\mathbf{v}_k$, $\mathfrak{h}_{(1)} = \mathfrak{h}_{S(1)}$
2) Define $\mathbf{g}_{(1)} = \mathfrak{h}_{(1)}$
Step 3. Semiorthogonal user selection.
While $i < N_s$, $i = i+1$

   a) $Q_i = \{\forall k \in Q_{i-1}, k \neq S(i-1) \mid \frac{\mathbf{g}_{(i-1)}^H \mathfrak{h}_k}{|\mathfrak{h}_k||\mathbf{g}_{(i-1)}|} < \alpha\}$

     If $|Q_i| \leq N_s - i + 1$, break While, end
   b) each user $k \in Q_i$ calculates $\mathbf{g}_k$, which is the component of $\mathfrak{h}_k$ that orthogonal to span$\{\mathbf{g}_{(1)}, \cdots, \mathbf{g}_{(i-1)}\}$, i.e.,

$$\mathbf{g}_{(k)} = \mathfrak{h}_k - \sum_{j=1}^{i-1} \frac{\mathbf{g}_{(j)}^H \mathfrak{h}_k}{|\mathbf{g}_{(j)}|^2} \mathbf{g}_{(j)}.$$

   c) $S(i) = \arg\max_{k \in Q_i} |\mathbf{g}_{(i)}|^2$, $\mathfrak{h}_{(i)} = \mathfrak{h}_{S(i)}$, $\mathbf{g}_{(i)} = \mathbf{g}_{S(i)}$.

$\mathfrak{h}_{(i)}$ and $\mathbf{g}_{(i)}$. Now, we explain **step 3**-(a). This step guarantees that the equivalent channels in Q_i are already semi-orthogonal to $\mathbf{g}_{(1)}, \cdots, \mathbf{g}_{(i-1)}$, which implies that $\mathfrak{h}_i \approx \mathbf{g}_i$ (α is a small positive constant). This algorithm enables secondary link to take advantages of multiuser diversity and leads to the establishment of **Theorem 4.1**, which is described as follows,

Theorem 4.1. *In the cognitive radio network where the primary link is a point-to-point MIMO channel and the secondary link is a multi-user uplink with K SUs, each SU is equipped with M_s antennas and PT, PR, and SR are equipped with M_p, N_p and N_s antennas, respectively. Based on the proposed two-stage user selection scheme, the sum rate of secondary link scales as $N_s \log\log K$ for K users with $K \to \infty$.*

The proof is in [20].

4.3.5 Simulation Results

Figure 4.3 shows the sum rate of secondary link as a function of $N_s \log(\log K)$ with different thresholds. It is assumed that PR, PT and SR are all equipped with four antennas, and each SU is equipped with three antennas. The number of dimensions on PR that are occupied by primary link is equal to 3, i.e., $s = M_s = 3$. In addition, we assume each SU has the same transmitting power, i.e., $\mathbf{P}_i = \mathbf{P}$. Hence, the threshold of each SU, γ_{th}, can be determined according to (4.10).

First of all, it is shown that higher threshold leads to higher average sum rate of secondary link. This is because a higher threshold results in more candidate users selected from **step 3** of **Algorithm 1**. As a result, the rate is boosted with multiuser

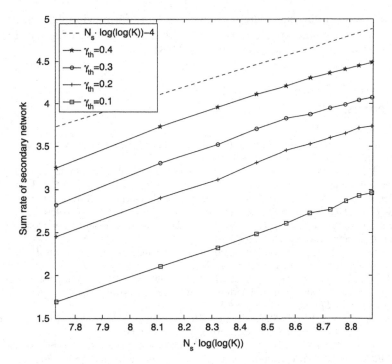

Fig. 4.3 The sum rate of secondary link as a function of $N_s \log(\log K)$, with different value of γ_{th}. $N_p = N_s = M_p = 4$, $s = M_s = 3$, $\alpha = 0.3$

diversity gain. Next, for the fixed threshold, we can see that with large K, the slope of each simulation curve almost matches the slope of $N_s \log(\log K)$, which confirms the results in **Theorem 4.1**.

4.4 Summary

In this Chapter, we studied the interference coordination techniques for cognitive radio networks. First, a technique was introduced to recycle spatial directions left unused by a primary MIMO link, so that they can be reused by secondary system, which is also a point-to-point MIMO link. We showed that the spatial white space of primary system can be exploited if the number of antennas at secondary transmitter is large enough. Then, in the case of cognitive uplink, we further investigated how to improve the performance of secondary system by relaxing the interference constraint of primary system and exploiting multiuser diversity. Specifically, a two-stage opportunistic user scheduling scheme has been proposed. In the first stage, the secondary users that cause interference leakage at the primary user less than a predetermined threshold are selected as candidate users. Then, N_s users should be selected among all the candidate users with an aim of achieving the highest rate. Analytical and simulation results both show that the sum rate of secondary link scales as $N_s \log(\log K)$ with sufficiently large K SUs.

References

1. Badoi, C., Prasad, N., Croitoru, V., & Prasad, R. (2011). 5G based on cognitive radio. *Wireless Personal Communications, 57*(3), 441–464.
2. Yang, L., Zhang, W., & Jin, S. (2015). Interferene alignment in device-to-device LAN underlaying cellular network. *IEEE Transactions on Wireless Communications, 14*(7), 3715–3723.
3. Guler, B., & Yener, A. (2014). Selective interference alignment fo MIMO cognitive femtocell networks. *IEEE Journal on Selected Areas in Communications, 32*(3), 439–450.
4. Mitola, J., & Maguire, G. (1999). Cognitive radio: Making software radios more personal. *IEEE Personal Communications, 6*, 13–18.
5. Haykin, S. (2005). Cognitive radio: Brain-empowered wireless communications. *IEEE Journal on Selected Areas in Communications, 23*(2), 201–220.
6. Perlaza, S. M., Fawaz, N., Lasaulce, S., & Debbah, M. (2010). From spectrum pooling to space pooling: Opportunistic interference alignment in MIMO cognitive networks. *IEEE Transactions on Signal Processing, 58*(7), 3728–3741.
7. Yang, L., & Zhang, W. (2015). On degrees of freedom region of three-user MIMO interference channels. *Transactions on Signal Processing, 63*(3), 590–603.
8. Yang, L., & Zhang, W. (2014). Interference alignment with asymmetric complex signaling on MIMO X channels. *IEEE Transactions on Communications, 62*(10), 3560–3570.
9. Zhang, R., & Liang, Y. (2008). Exploiting multi-Antennas for opportunistic spectrum sharing in cognitive radio networks. *IEEE Journal of Selected Topics in Signal Processing, 2*(1), 88–102.
10. Telatar, E. (1999). Capacity of multi-antenna Gaussian channels. *European Transactions on Telecommunications, 10*(6), 585–596.
11. Amir, M., El-Keyi, A., & Nafle, M. (2010). Opportunistic interference alignment for multiuser cognitive radio. In *Proceedings of the IEEE Information Theory Workshop*, Cairo, 6–8 Jan 2010.
12. Shen, C., & Fitz, M. P. (2011). Opportunistic spatial orthogonalization and its application in fading cognitive radio networks. *IEEE Journal on Selected Topics in Signal Processing, 5*(1), 182–189.
13. Zhang, L., Liang, Y. C., & Xin, Y. (2008). Joint beamforming and power allocation for multiple access channels in cognitive radio networks. *IEEE Journal on Selected Areas in Communications, 26*(1), 38–51.
14. Ban, T., Choi, W., Jung, B. C., & Sung, D. K. (2009). Multi-user diversity in a spectrum sharing system. *IEEE Transactions on Wireless Communication, 8*(1), 102–106.
15. Zhang, R., & Liang, Y. C. (2010). Investigation on multiuser diversity in spectrum sharing based cognitive radio networks. *IEEE Communications Letters, 14*(2), 133–135.
16. Yang, L., & Zhang, W. (2013). Opportunistic interference alignment in heterogeneous two-cell uplink network. In *Proceedings of the IEEE Interantional Conference on Communications (ICC 2013)*, Budapest, pp. 5448–5452, 9–13 June 2013.
17. Hamdi, K., Zhang, W., & Letaief, K. B. (2009). Opportunistic spectrum sharing in cognitive MIMO wireless networks. *IEEE Transactions on Wireless Communication, 8*(8), 4098–4109.
18. Yoo, T., & Goldsmith, A. (2006). On the optimality of multiantenna broadcast scheduling using zero-forcing beamforming. *IEEE Journal on Selected Areas in Communications, 24*(3), 528–541.

19. Zhang, W., & Letaief, K. B. (2008). Opportunistic relaying for dual-hop wireless MIMO channels. In *Proceedings of the IEEE Global Communications Conference (GLOBECOM 2008)*, New Orleans, 30 Nov–4 Dec 2008.
20. Yang, L., Zhang, W., Zheng, N., & Ching, P. C. (2014). Opportunistic user scheduling in MIMO cognitive radio networks. In *Proceedings of the IEEE International Conference on Acoustics, Speech and Signal Processing (ICASSP 2014)*, Florence, pp. 7303–7307, 4–9 May 2014.

Chapter 5
Conclusions

The fifth generation (5G) cellular wireless network is seen as a promising next generation cellular technology. It is expected to be a mixture of networks of various tiers, various backhaul connections, and different radio access technologies that are accessed by a great number of wireless devices. However, due to the heterogeneity and dense deployment of wireless devices and base stations, the interference will become increasingly intensive, and the performance of 5G networks will be seriously degraded if the interference is not well managed.

In this Brief, we have discussed various interference coordination techniques in 5G cellular network. First, we addressed the problem of inter-cell interference, which has huge impact on the performance of cell-edge users. In order to coordinate the interference effectively, a certain level of cooperation between base stations is needed. With CSI-sharing cooperation level, a beamforming scheme was proposed to coordinate the interference among the users. Based on the scheme, optimal degrees of freedom region can be achieved. With virtual MIMO cooperation level, the concept of cloud radio access network was introduced, which can exploit the interference links for the transmission of useful signals. Secondly, we considered inter-tier interference between cellular networks and D2D networks. As one of the key technologies of 5G network, D2D communications brings many potential benefits. However, it also brings new challenges in interference management, i.e., not only the interference between two networks should be taken care of, but also the interference among D2D users should be considered. Two interference coordination schemes were proposed for a typical D2D LAN underlaying cellular network. Performance analysis showed that based on the proposed schemes, the interference generated on the cellular links is eliminated or well controlled, while the quality of service of the D2D LAN can also be guaranteed. Finally, we focused on cognitive radio, which is another key technique for 5G. A user scheduling algorithm was utilized to improve the performance of both primary and secondary systems by taking advantage of multiuser diversity.

© The Author(s) 2015
L. Yang, W. Zhang, *Interference Coordination for 5G Cellular Networks*,
SpringerBriefs in Electrical and Computer Engineering,
DOI 10.1007/978-3-319-24723-6_5

Next, we discuss some other future directions of cellular networks. To achieve dramatic gains and simplify the required signal processing, massive MIMO system or large-scale antenna systems have been widely discussed in these days, where each base station is equipped with orders of magnitude more antennas, e.g., 100 or more [1–4]. The main advantage is that with large number of antennas at base station, the channel vectors of different users become asymptotically orthogonal, which means the effect of small-scale fading channels is eliminated [3, 4]. On the other hand, the implementation of massive MIMO is also facing many challenges such as pilot contamination, training overhead, and hardware cost, etc. [5, 6]. Another trend in 5G networks is the utilization of much wider range of frequency spectrum. Currently, most wireless systems operate in the frequency range that extends from several hundred MHz to a few GHz. However, these bands have been almost fully occupied. A solution for this spectrum crisis is to exploit the vast amounts of relatively idle spectrum that exists in the range of 30–300 GHz, where the wavelengths are 1–10 millimeters (mm). The communications over this spectrum is referred to as 'mmWave' communication [7–9]. Hence, it is in general best suitable for short range communications [9, 10].

References

1. Lu, L., Li, Y., Swindlehurst, A., Ashikhmin, A., & Zhang, R. (2014). An overview of massive MIMO: Benefits and challenges. *IEEE Journal on Selected Topics in Signal Processing, 8*(5), 742–758.
2. Rusek, F., Persson, D., Lau, B. K., Larsson, E. G., Marzetta, T., Edfors, O., & Tufvesson, F. (2013). Scaling up MIMO: Opportunities and challenges with very large arrays. *IEEE Signal Processing Magazine, 30*(1), 40–46.
3. Ngo, H., Larsson, G., & Marzetta, T. (2013). Energy and spectral efficiency of very large multiuser MIMO systems. *IEEE Transactions on Communications, 61*(4), 1436–1449.
4. Hoydis, J., Brink, S., & Debbah, M. (2013). Massive MIMO in the UL/DL of cellular networks: How many antennas do we need? *IEEE Journal on Selected Areas in Communications, 31*(2), 160–171.
5. Yin, H., Gesbert, D., Filippou, M., & Liu, Y. (2013). A coordinated approach to channel estimation in large-scale multiple-antenna systems. *IEEE Journal on Selected Areas in Communications, 31*(2), 264–273.
6. Zeng, Y., Zhang, R., & Chen, Z. (2014). Electromagnetic lens-focusing antenna enabled massive MIMO: Performance improvement and cost reduction. *IEEE Journal on Selected Areas in Communications, 32*(6), 1194–1206.
7. Rangan, S., Rappaport, T. S., & Erkip, E. (2014). Millimeter-wave cellular wireless networks: Potentials and challenges. *Proceedings of IEEE, 102*(3), 366–385.
8. Rappaport, T. S., Heath, R. W., Jr., Daniels, R. C., & Murdock, J. N. (2014). *Millimeter wave wireless communications*. Upper Saddle River: NJ, Prentice Hall.
9. Alkhateeb, A., Mo, J., Prelcic, G., & Heath, R., Jr. (2014). MIMO precoding and combining solutions for millimeter wave systems. *IEEE Communications Magazine, 52*(12), 122–131.
10. Bai, T., Alkhateeb, A., & Heath, R., Jr. (2014). Coverage and capacity of millimeter wave cellular networks. *IEEE Communications Magazine, 52*(9), 70–77.

Printed in the United States
By Bookmasters